Bryozoan Evolution

TITLES OF RELATED INTEREST

A biologist's advanced mathematics
D. Causton

Cell movement and cell behaviour
J. M. Lackie

Echinoid palaeobiology
A. Smith

Eukaryote genome in development and evolution
B. John & G. Miklos

Invertebrate palaeontology and evolution
E. N. K. Clarkson

Microfossils
M. D. Brasier

A natural history of Nautilus
P. Ward

Paleopalynology
A. Traverse

A practical approach to sedimentology
R. Lindholm

Quaternary paleoclimatology
R. Bradley

Rates of evolution
P. S. W. Campbell & M. F. Day

Bryozoan Evolution

F. K. McKinney
Department of Geology, Appalachian State University

and

J. B. C. Jackson
Smithsonian Tropical Research Institute

Boston
UNWIN HYMAN
London   Sydney   Wellington

Unwin Hyman, Inc.
8 Winchester Place, Winchester, Mass. 01890, USA

Published by the Academic Division of
Unwin Hyman Ltd
15/17 Broadwick Street, London W1V 1FP, UK

Allen & Unwin (Australia) Ltd,
8 Napier Street, North Sydney, NSW 2060, Australia

Allen & Unwin (New Zealand) Ltd in association with the
Port Nicholson Press Ltd,
60 Cambridge Terrace, Wellington, New Zealand

First published in 1989

Library of Congress Cataloging in Publication Data

McKinney, Frank K. (Frank Kenneth)
   Bryozoan evolution.
(Special topics in paleontology, ISSN 0261–0515 ; 2)
Bibliography: p.
Includes index.
1. Bryozoa, Fossil.   2. Evolution.   I. Jackson,
Jeremy B. C., 1942–   .   II. Title.   ~~III. Series.~~
QE798.M35   1987   564′.7   87–14580
ISBN 0–04–560012–0

British Library Cataloguing in Publication Data

McKinney, F. K.
   Bryozoan evolution. —— (special topics in
palaeontology, ISSN 0261–0515 ; 2).
1. Bryozoa
I. Title   II. Jackson, Jeremy B.C.
III. Series
594′.7   QL396
ISBN 0–04–560012–0

Typeset in 10 on 12 point Times by Nene Phototypesetters Ltd
and printed in Great Britain by Biddles of Guildford

Preface

One of the fundamental questions of paleobiology is the extent to which the major features of morphological evolution are shaped through adaptation by natural selection, through developmental constraints imposed on taxa by their basic design, or merely by chance events. The history of evolutionary science is fraught by pendulum swings between these alternatives, even though it is almost certainly true that each process has an important role. The currently growing trend to discount adaptationist interpretations is salutary in revealing fuzzy thinking of the "everything must have a purpose and be the best for the job" sort, but in their zeal some followers of this revived creed seem all too willing to throw out results of much good functional analysis as well. In fact, we have few data on which to base an opinion. Scattered bits of evidence from one taxon or another, selected for whatever reason, do not constitute the stuff of objective comparative analysis. What we need instead is to examine systematically the evolutionary histories of several higher taxa, identify prominent patterns or trends, and try to understand their underlying basis from all perspectives. We hope that this book contributes in some way towards that goal.

Bryozoans are an excellent group for the study of evolution. They are among the three dominant groups of Paleozoic fossils and are abundant from the Ordovician to Recent. Bryozoans are also morphologically complex, with many of their functional attributes well represented in their skeletal structure. Most are sessile and commonly die in such a way that their relations to one another are preserved as they were in life. Preservation of colony shape and budding, patterns of previous interaction, and life history offer some of the best evidence of ancient natural selection that fossils provide. Lastly, bryozoans are the only major phylum of exclusively clonal animals and, with the exception of a few bizarre species, they are all colonial as well. This modular construction allows two different levels of morphological analysis: that of the building blocks – the zooids – and of the budding and packaging of zooids to form differently shaped colonies. This book concentrates on major patterns of evolution of bryozoan colony form, rather than on zooids, because we understand the colonies better, and it is the colony (or clone) that constitutes the genetic individual comparable to a single clam or snail.

The book begins with an introduction to bryozoan structure, function, classification, and the state of bryozoan species level taxonomy and its bearing on described phylogenies. Chapter 4 outlines our fundamental premise that growth pattern and form of a bryozoan colony is an expression of its ecological niche more than its phylogenetic history. Repetition of similar forms in similar situations throughout the history of the phylum defines the major themes of bryozoan evolution which follow. These, in order, are patterns of life history, feeding, and varying solutions to problems related to encrusting, erect, and free-living or rooted life habits. We conclude with a discussion of several long-term, apparently progressive and adaptive trends in the evolution of colony form.

This book is an outgrowth of our mutual interests in the biology and paleobiology of bryozoans. It contains those topics with which we are most familiar and which we believe have been most important in molding the history of the phylum. We have tried to approach different questions consistently and have had to generate much new information in the process, particularly in Chapters 4, 5, 6, and 8. We hope that because of our very different backgrounds we bring a broader perspective to bryozoan evolution than could either of us working alone. Certainly, we have learned enormously from one another.

The late Tom Schopf first suggested we write this book. We are grateful for his interest and support. Most of the ideas and much of the manuscript have benefitted from the advice and criticism of Richard Boardman, Leo Buss, Alan Cheetham, Pat Cook, Karl Kaufmann, Nancy Knowlton, Scott Lidgard, Marg McKinney, Paul Taylor, William Turpin and Judith Winston.

To Marg and Nancy

Contents

Preface   *page*   vii
List of tables   xiii

1   Bryozoans as modular machines   1
    1.1   Modular construction   1
    1.2   Integration of modules   10
    1.3   Major taxa of marine Bryozoa   14

2   Species relationships and evolution   32
    2.1   Morphological characterization of species   32
    2.2   Is morphology adequate?   36
    2.3   Inferred species evolution and lineages   41
    2.4   Untested phylogenies   51

3   Growth and form   52
    3.1   Encrusting growth   52
    3.2   Erect growth   55
    3.3   Free-living growth   68
    3.4   Rooted growth   69

4   Growth forms as adaptive strategies   72
    4.1   Frequencies of growth forms among major taxa   72
    4.2   Growth form model   74
    4.3   Zooidal characteristics and colony form   76
    4.4   Environmental distributions   80
    4.5   Growth forms in the fossil record   89
    4.6   Growth forms as adaptive strategies   94

5   Bryozoan life histories   97
    5.1   Reproductive ecology   97
    5.2   Six case studies   100
    5.3   Life history patterns   112
    5.4   Size and age   114
    5.5   Dispersal   115
    5.6   Heterogeneity within colonies   117

6   Feeding: a major sculptor of bryozoan form   119

    6.1   Food   119
    6.2   Feeding structures   121
    6.3   Feeding currents and behavior   128
    6.4   Feeding and colony form   135

7   Encrusting growth: the importance of biological
    interactions   145

    7.1   Short-lived or unstable substrata   146
    7.2   Substrata of intermediate longevity and stability   149
    7.3   Long-lived, stable substrata   151
    7.4   Settlement panels   153
    7.5   Characteristics of abundant encrusting bryozoans on
         stable substrata   154
    7.6   Paleoecology of encrusting bryozoans   159

8   Erect growth: problems of breakage and flow   167

    8.1   Effects of erect growth   168
    8.2   Flexible erect growth   171
    8.3   Rigidly erect growth   173

9   Life on and in sediments: problems of substratum
    stability   190

    9.1   Free-living bryozoans   190
    9.2   Rooted bryozoans   203
    9.3   Interstitial bryozoans   206
    9.4   Major evolutionary trends   206

10   Trends in bryozoan evolution   209

    10.1   Trends in the design, distribution, and relative
          frequency of growth forms   209
    10.2   Interpretation of trends   211

References   213
Index   233

List of tables

1.1 Characteristics of the marine Bryozoa

4.1 Variation in zooidal size, elongation, spacing, and incidence of polymorphic zooids in relation to colony form among recent and early Tertiary (Paleocene, Eocene) cheilostomes.

4.2 Taxonomic distributions of species reported from brackish water world-wide, and for all species from all environments in our compilation of Atlantic and western Mediterranean bryozoans

5.1 Outcome of interspecific overgrowth interactions involving younger (distal) and older (proximal) areas of *Steginoporella* sp. colonies

5.2 Differences between bryozoans under corals and those on settlement panels after 3–4 months at 10 and 20 m depths on north-coast Jamaican reefs

5.3 Comparison of geographic distributions of tropical and subtropical western Atlantic cheilostomes with cyphonautes versus brooded larval development

6.1 Sizes of feeding structures in Recent marine bryozoans

6.2 Estimates of mouth diameter and lophophore diameter of some fossil bryozoans

6.3 Degree of physiological and structural integration in gymnolaemates with different colony current patterns

7.1 Ecological characteristics of the principal groups of encrusting epifauna on shells of *Pinna bicolor* and on pier pilings at Edithburgh, South Australia

7.2 Budding characteristics of the most abundant species of encrusting bryozoans on ephemeral and stable substrata

8.1 Averages for values of branch thickness for Recent, Oligocene, and Paleocene species of adeoniform cheilostomes that exhibit proximal thickening

10.1 Temporal trends in bryozoan evolution described in this book

1 Bryozoans as modular machines

Bryozoans are colonies of modular units termed **zooids**. The shape and functioning of the colony depends on how the modules are assembled and the kinds of modules present. Colonial growth potentially enables bryozoans to develop unrestricted variations in form, but only a relatively small number of basic growth forms have ever developed and, indeed, have commonly reappeared many times throughout the history of the phylum. This repeated evolution and stability of growth forms, the improvement of design within them, and their distributions in space and time are the major subjects of this book. In contrast, the evolutionary history of individual modules has witnessed profound changes among the different groups of Bryozoa. These changes are more important to the ways that parts of colonies function and are held together than to their form *per se*. They are also of fundamental systematic and evolutionary significance, because zooids provide consistently more complex and stable taxonomic characters than colony form. In this chapter we describe characteristics of zooids essential for understanding how bryozoan colonies work. We then introduce the major taxa of marine bryozoans and outline their geological histories.

1.1 Modular construction

Zooids are morphologically discrete functional units (Fig. 1.1) analogous to entire unitary animals, but varyingly interdependent for the function and survival of an entire bryozoan organism. Zooids are bounded by walls of tissue and an outer noncellular organic membrane (**cuticle**) that enclose a coelomic space. Within this space are tissues that are varyingly organized according to particular zooidal functions. Zooids in virtually all species are physiologically connected through pores in the intervening walls, or through confluent space between outer edges of walls and cuticle, or directly where no walls intervene. Zooids are minute, in no case greater than $1\,mm^3$, and typically much less. The longest reach several millimetres, but these generally have diameters of about 0.2 mm. Sexually mature colonies of some species may contain only two or three zooids, but numbers more typically range from tens to thousands and may exceed 10 million. Numbers of zooids within colonies and the area they cover may vary, therefore, over seven orders of magnitude.

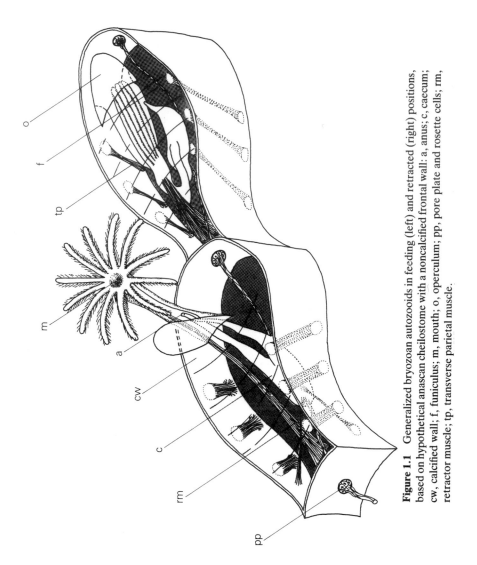

Figure 1.1 Generalized bryozoan autozooids in feeding (left) and retracted (right) positions, based on hypothetical anascan cheilostome with a noncalcified frontal wall: a, anus; c, caecum; cw, calcified wall; f, funiculus; m, mouth; o, operculum; pp, pore plate and rosette cells; rm, retractor muscle; tp, transverse parietal muscle.

1.1.1 The basic modules: autozooids

An **autozooid** (Fig. 1.1) is composed of a feeding unit, which is the **polypide**, and an enclosing sheath of tissue from which the tentaculate end of the polypide protrudes for feeding. Epithelium within the enclosing sheath of tissue may secrete calcium carbonate that in many marine species forms a rigid zooidal skeleton, termed the **zooecium**. The **tentacle sheath** is the part of the enclosing tissue that attaches to the base of the tentacles, surrounding them when the polypide is retracted and being everted when the polypide is protruded.

Autozooidal polypides consist of a U-shaped gut and a ring of tentacles termed the **lophophore**. When protruded to the feeding position, the tentacles spread outward from their bases, forming a funnel or bell that is centered on the mouth. Each tentacle contains coelomic space (that interconnects with others through a coelomic ring at the base of the lophophore), a hollow collagenous supporting rod, muscles, and other tissues (Lutaud 1955, 1973, Smith 1973, Gordon 1974). Three compound tracts of cilia line each tentacle. One tract is inner, or frontal, and passes captured particles toward the mouth. The other two are laterally placed on opposite sides of the tentacle, and their coordinated beating produces waves that pass along each side, thereby driving water from the area above the lophophore, through the spaces between tentacles, to the region below the lophophore. Food particles are captured by a variety of behaviors, the most widespread of which involves localized reversal of ciliary beating that traps and propels particles around tentacles and down towards the mouth (Strathmann 1973, 1982). Alternatively, the tentacle rapidly bends inward, carrying the particle towards the mouth.

In colonies in which zooids are monomorphic, all are autozooids and all or some are engaged in reproduction. In others, reproductive zooids have distinctive morphologies, and various other specialized modules may also occur. However, all colonies must have at least some feeding autozooids to provide nutrients for growth and functioning of the specialists.

1.1.2 Specialized modules: heterozooids

Specialization of function has produced widespread zooidal polymorphism in bryozoans (Fig. 1.2, Silén 1977, Boardman *et al.* 1983). The evolution of modular specialization within bryozoan colonies and of increased precision of placement of the specialists has been a common theme of evolution in the phylum. It characterizes especially the gymnolaemate order Cheilostomata (§ 1.3.2). Polymorphs may be divided into autozooids and, in contrast, various nonfeeding **heterozooids**, which are present in many species and occur as one or more specialized types within colonies. Functions performed by heterozooids include protection, reproduction, locomotion, plumbing,

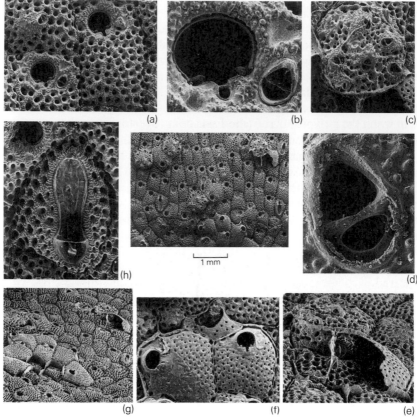

Figure 1.2 Polymorphism within a 10 mm² area of a frontally budding colony of *Stylopoma spongites* shown at the center. (a) Two autozooids with two types of adventitious avicularia, one to the right of the orifice and the second on a prominence of its own construction at the proximal end of the neighboring zooid. (b) Close-up of orifice and associated avicularium. (c) Giant ovicell supporting at least four adventitious avicularia. (d) Close-up of one of the avicularia on ovicell. (e) Low-angle view of the same ovicell (top) with beginnings of two frontally budded zooids adjacent. (f) Group of newly frontally budded zooids. (g) Low-angle view showing elevation of the same two regions of frontal budding. (h) Large vicarious avicularium.

structural support, and many more. The function of many heterozooids is still unknown.

Commonly, eggs are brooded in morphologically distinct female heterozooids. In many Cheilostomata, however, brooding occurs in chambers termed **ovicells** which are located distal to the orifice of female or hermaphroditic autozooids, lie above the proximal end of the next zooid, or are imbedded in the proximal end of the next zooid. Some ovicells may themselves be heterozooids or heterozooidal complexes (Woollacott & Zimmer 1972, Soule 1973), each associated with a female autozooid, so that larval production is a cooperative effort between two or more zooids. In members of the class Stenolaemata (§ 1.3.1), female heterozooids have inflated brood chambers below their apertures. These brood chambers may ramify out into the space between adjacent zooids, just below the level of their apertures (see Fig. 1.5b). Brood chambers in some stenolaemates are apparently not associated with one identifiable maternal zooid but may communicate directly with several contiguous zooids (Borg 1933, Boardman 1983).

Male heterozooids are uncommon, and sperm production generally occurs in some or all autozooids (Silén 1966). Mature sperm, which originate from testes on a tissue strand (the **funiculus**) extending from the gut to the body wall, are released from tips of tentacles. Male heterozooids have a smaller number of tentacles (which are unciliated) than do the associated autozooids (Cook 1968, Gordon 1968, Ryland 1979a).

Avicularia (Fig. 1.3) are variously shaped heterozooids that are widespread in the cheilostomes and also evolved in a Mesozoic clade of stenolaemates (Taylor 1985b). They are based on modification of the orifice

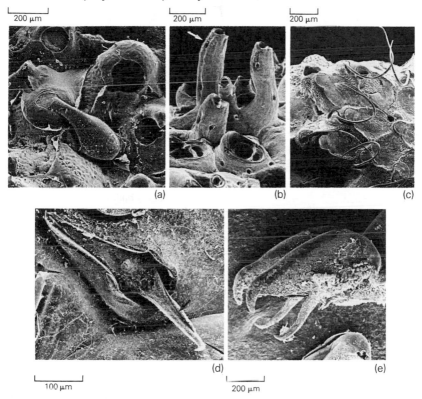

Figure 1.3 Various avicularia. (a) Interzooidal avicularium with spatulate mandible in *Celleporaria albirostris*. (b) Tubular adventitious avicularia (arrowed) and interzooidal avicularia (asterisk) in *Orthoporidra compacta*. (c) Vibracula in *Discoporella umbellata depressa*. (d) Adventitious avicularium with open mandible (arrowed) in *Hippopodina feegeensis*. (e) Twinned pedunculate avicularium in *Beania magellanica*. (Photographs courtesy of J. E. Winston.)

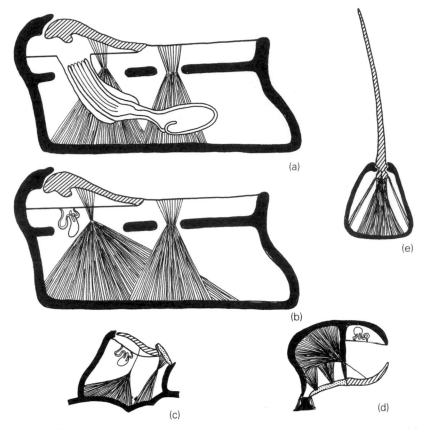

Figure 1.4 Types and structure of avicularia: (a) vicarious or intercalated with functional polypide; (b) the same with rudimentary polypide; (c) adventitious sedentary; (d) adventitious pedunculate; (e) vibraculum. Only skeleton, membranes, muscles and polypides are shown. Homologous structures are indicated by similar patterns, the most critical of which are those derived from the autozooidal operculum (diagonally lined). (Figs. (b) and (d) modified from Silén 1977, courtesy of Academic Press; (c) and (e) after Marcus 1939.)

and operculum, with reduction of the polypide to a sensory or rudimentary organ (Fig. 1.4). Typically, the orifice is elongated and narrowed distally, and the operculum, termed the **mandible** in avicularia, is jaw-shaped, with the distal end generally recurved as a hook and closing inside the elevated rim of the orifice. The orifice is opened when abductor muscles along the proximal portion of the mandible contract, pulling the proximal end down so that beyond the fulcrum points the distal portion rotates away from the orifice, and in some forms opens more than 180°. The mandible may be snapped shut extremely rapidly by adductor muscles attached to one or two ligaments that are inserted about midway along its length.

Avicularia vary in position and size. The largest are **vicarious** (see Fig. 1.2h), occupying the position of autozooids; others are **interzooidal** (see Fig.

1.3a), occurring in spaces between autozooids but not substituting for them. In contrast, generally small **adventitious** avicularia occur on the exposed surfaces of autozooids and heterozooids (see Fig. 1.2a–d). These latter avicularia may have their orifices flush with the surface of the supporting zooid, or the margin may slope from low at their base to highly elevated at their tips. Some adventitious avicularia are lightly calcified and are attached by a short peduncle, with the larger part of the zooid, the **rostrum**, overarching the mandible when the orifice is closed (see Figs. 1.3e, 1.4d). It is these pedunculate forms that give avicularia their name, because "they curiously resemble the head and beak of a vulture in miniature, seated on a neck and capable of movement, as is likewise the lower jaw or mandible" (Darwin 1872).

Function of avicularia has been the subject of much speculation (Ch. 10, Winston 1984b). Their extreme morphological variability, even between closely related species (see Fig. 2.1), and the occurrence of more than one form within colonies of a species argue for a diversity of roles. A study involving both observation of live specimens and structural analysis suggests that pedunculate avicularia in two species of *Bugula* defend against, or at least deter, predators such as small annelids, amphipods, and pycnogonids that have graspable body parts of about 0.05 mm diameter or less (Kaufmann 1971). In contrast, Winston (1984b) has suggested the possibility of a structural origin for avicularia that is independent of any functional cause.

Vibracula (see Figs. 1.3c, 1.4e) are a type of avicularia in which the opercula are hypertrophied into long setae (Silén 1938, 1977) that can sweep round, rotating against the frontal membrane. These are traditionally interpreted as having a cleaning function because of their ability to sweep across the colony surface, but they also provide locomotion for some cap-shaped, free-living cheilostomes (Ch. 9; Cook & Chimonides 1978).

Small heterozooids with tiny orifices and diminutive single-tentacled polypides, **nanozooids**, occur in a few living stenolaemates (Fig. 1.5a, b). Some nanozooids are primary, that is, budded in a regular pattern among the autozooids, whereas others follow autozooids in a zooecium (an example of **intrazooidal polymorphism**) with an inverted, funnel-shaped terminal diaphragm that reduces the orifice to the diameter of a single tentacle. The tentacle of a primary nanozooid sweeps around across the surrounding surface and perhaps has a cleaning function (Silén & Harmelin 1974). Function of the secondary nanozooids is unknown. Secondary nanozooids have a long fossil record, occurring in older or more heavily calcified parts of colonies of some Paleozoic fenestrate stenolaemates (McKinney 1977b, Bancroft 1986).

In both gymnolaemates and stenolaemates, a large number of zooids of different structure that are known or suspected to lack polypide and orifice are collectively termed **kenozooids**, though a plethora of terms is applied to particular types. They occur intercalated among other zooids, along colony margins, or fill large surface areas (Fig. 1.6b). Kenozooids are presumed

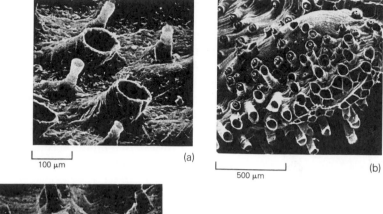

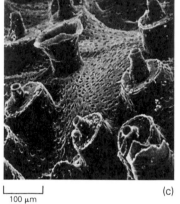

Figure 1.5 Nanozooids in living cyclostomes. (a) Primary nanozooids budded alternately with autozooids in *Diplosolen obelium*. (b) and (c) Secondary nanozooids developed from autozooids after the end of the peristome breaks off in *Plagioecia dorsalis*. Also note ovicell (○) at top right of (b). (From Harmelin 1976, courtesy Musée Océanographique, Monaco.)

Figure 1.6 (a) Colony and (b) helically twisted bilaminate branch of *Spiralaria florea*, composed of kenozooids (k), autozooids (a) and adventitious avicularia (v). (From McKinney & Wass 1981, courtesy of Olsen & Olsen.)

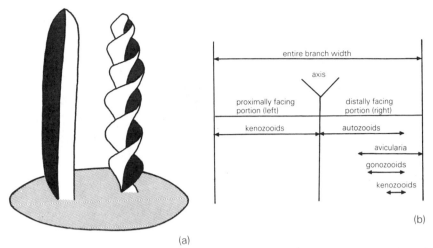

Figure 1.7 (a) Branch twisting and (b) distribution of polymorphic zooids across one face of a branch of the bilaminate cheilostome *Spiralaria florea*. The opposite face is similarly organized. (From McKinney & Wass 1981, courtesy of Olsen & Olsen.)

to have various strengthening, attachment, supportive, and space-filling functions.

Even the spines on the frontal surface of cheilostome zooids may be heterozooids. Including spines, avicularia, and ovicells, an apparently "individual" cheilostome autozooid may comprise an association of upwards of a dozen zooidal members.

Heterozooids may be distributed through a colony in regular fashion or may appear sporadically. An extraordinary example of highly ordered distribution is *Spiralaria florea* (Fig. 1.6). In this species, each bilaminate branch is twisted like a spiral piece of pasta so that half of each surface (of what one might imagine as the original surface of a ribbon) faces distally and the other half faces proximally (Fig. 1.7). The proximally facing portion of each face is occupied entirely by kenozooids. In contrast, on the distally facing portions there are autozooids throughout, with outwardly enlarging adventitious avicularia in the marginal quarter of the ribbon, and kenozooids smoothing the margin of the branch.

Other patterns include the proliferation of kenozooids in regions of crowding (Fig. 1.8) or of marked change in growth direction in response to contacts between colonies (Buss 1981b). Frontal spines may develop in response to grazing by predators or as a defense against overgrowth (§ 7.5).

1.1.3 Nonmodular colony regions

The colonies of many marine bryozoans have areas that are not part of any autozooid or heterozooid. Such areas are referred to as **extrazooidal**. They are solid skeleton or a series of skeletal cysts that serve to fill space or strengthen the surface of a colony (see Figs. 4.7 & 8.8).

2 mm

Figure 1.8 Uniformly sized, rectangular autozooids in uncrowded regions, and variably shaped and sized kenozooids in crowded portions of *Membranipora membranacea*. (From Stebbing 1973b, courtesy of Academic Press.)

1.2   Integration of modules

Colonial integration is a measure of the degree of interdependence and cooperation among modules within a colony. Important aspects of integration include: increases in colonial rather than modular functions, the development of nonfeeding heterozooids, and the ability of a colony to develop heterozooids as an intrinsic part of the colony structure or in response to local environmental conditions. Increased colonial integration might be advantageous in several ways, including the production of stronger and more coordinated feeding currents, rapid directional growth, more useful distribution of polymorphic zooids, and superior architectural design, all of which may increase fitness.

Boardman & Cheetham (1973) recognized six measures of integration in bryozoans:

(a) type of zooidal walls,
(b) type of interzooidal connections by zooidal soft tissues,
(c) timing and source of development of extrazooidal tissues,
(d) degree of morphological differences between generations of zooids,
(e) degree of polymorphism, and
(f) position of heterozooids.

The degree of colonial integration varies widely in marine bryozoans, from some Paleozoic encrusting stenolaemates in which monomorphic zooids had no soft tissue connections and apparently functioned independently, to highly integrated free-living cheilostomes that have precisely positioned heterozooids, extensive extrazooidal tissues, highly determinate form, and corporate colony behavior (§ 9.1).

Cheilostomes exhibit a striking progressive increase in integration through the history of the order, despite the continued existence of poorly integrated species (Fig. 1.9). In contrast, there has been no net change among stenolaemates taken as a whole (Boardman & Cheetham 1973), although some clades within the class, such as the fenestrates discussed in Section 8.3, did develop increased integration through time.

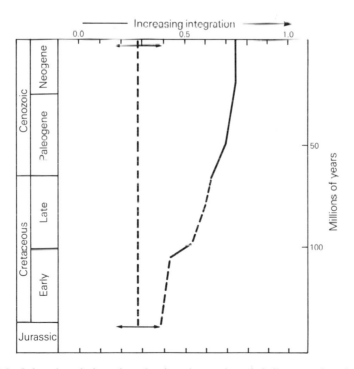

Figure 1.9 Inferred evolution of total colony integration of cheilostomes based on three structural states and three functional states, with 0.0 equal to complete lack of integration and 1.0 equal to full integration. (From Boardman & Cheetham 1973, courtesy of the authors.)

Figure 1.10 Diagrams of stenolaemates in which walls (black) are breached by interzooidal pores. Nutrient transfer may occur only through pores in calcified walls in *Crisia* (a) and *Tubulipora* (b), but in *Lichenopora* (c) nutrients may pass through the space between outer edges of calcified walls and the outer cuticle (stippled), as well as through pores in the walls. (After Neilsen & Pedersen 1979, courtesy of The Royal Swedish Academy of Sciences.)

1.2.1   Feeding and integration

The currents generated by cilia of individual lophophores may combine in local regions to produce coherent flow patterns over or past some portion of a colony's surface (Winston 1978, 1979). Such cooperative behavior is a result of and reflects colony integration. The translation of nutrients from feeding zooids to nonfeeding portions of colonies, such as zones of growth and, in the most integrated forms, nonfeeding heterozooids, is another aspect of integrated colonial functions.

Some degree of colonial integration is necessary for transfer of nutrients from feeding zooids to other parts of the colony, including the growing margins where there are no functional polypides. In stenolaemates, transfer of nutrients occurs within the space between the outer cuticle and the outer surface of the interior skeleton, through pores in zooidal walls, or both (Fig. 1.10). Most stenolaemates have only one of these two pathways available. The least integrated stenolaemates lack pores or have only a single pore for interzooidal communication and are restricted to poorly organized, runner-like encrusting growth. When interzooidal communication is more extensive, a diversity of growth forms is possible.

Nutrient transfer to nonfeeding areas in gymnolaemates occurs via the funiculus, a tissue strand perhaps homologous to blood vessels (Carle & Ruppert 1983, Lutaud 1985) that interconnects zooids by pinched "cell

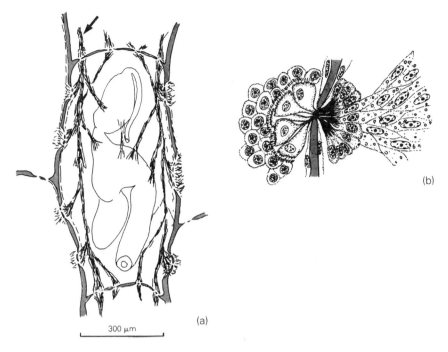

(b)

(a)

300 μm

Figure 1.11 (a) Funicular system (arrowed) attached to polypide and interconnected through pore plates in zooidal walls, and (b) detail of tissue-plugged pore in pore plate, with polarity of nutrient flow toward funicular strand on right. (Fig. (a) after Banta 1969, courtesy of Alan R. Liss; (b) after Bobin 1977, courtesy of Academic Press.)

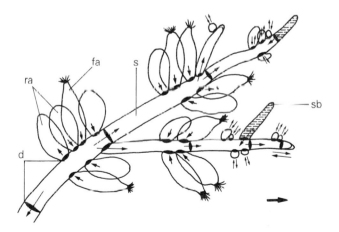

Figure 1.12 Flow of nutrients along funiculus and through pore rosettes from feeding autozooids, through stolonal kenozooids, to regions of growth in *Bowerbankia imbricata*; orientation of cells indicates that stolon buds filled by dashes are no longer growing: d, diaphragms; fa, functional autozooid; ra, regressed autozooids; s, stolon; sb, stolonal bud. (From Bobin 1977, courtesy of Academic Press.)

rosettes" that extend through pore plates in zooidal walls (Fig. 1.11, Bobin 1977, Lutaud 1983). The funiculus has a branch that connects with the stomach in autozooids and apparently serves as the path for streaming nutrients, especially lipids, that may be stored as reserves within the autozooid or that may move through rosettes in pore plates toward specific regions, generally distal (Fig. 1.12).

Transfer of nutrients from feeding autozooids to growing margins is accomplished by regional polarity of flow within a colony (Best & Thorpe 1985). If the polarity of nutrient flow throughout the colony is towards specific areas, directional growth occurs. A good example of such directional growth may be seen in the cheilostome *Membranipora membranacea* on fronds of the alga *Laminaria*, where the colony grows rapidly upcurrent towards the younger parts of the frond, and therefore away from the flapping end of the algal stipe where abrasion and fragmentation are more active (Ryland & Stebbing 1971).

1.2.2   Colony form and integration

Colony form is correlated with colony integration (Fig. 1.13). Encrusting bryozoans in general show poor structural integration and vary from poor to good in their functional integration. In contrast, rigidly erect and free-living forms exhibit high levels of structural and functional integration (§ 4.3).

1.3   Major taxa of marine Bryozoa

Two of the three classes of Bryozoa are marine: the Stenolaemata are exclusively marine and the Gymnolaemata are dominantly so. The gymnolaemates are today by far the most abundant and diverse class and have been since the Late Cretaceous; before that the stenolaemates dominated (Fig. 1.14). The class Phylactolaemata occurs only in freshwater, is unmineralized, and has left a paltry record of cyst-like vegetative reproductive structures. Phylactolaemates are not included in this book.

1.3.1   Stenolaemata

Stenolaemates have calcified tubular or sac-shaped zooids whose principal axis lies at an acute angle to the local direction of growth (**distal** direction) of the colony (Fig. 1.15) A **membranous sac** surrounds each polypide. Except for the terminal membrane that extends across the skeletal aperture, and interzooidal pores within two suborders, bounding walls of stenolaemate zooids are completely calcified. The terminal membrane folds down around the orifice, extends into the zooid and continues as a vestibular wall that recurves and attaches by a ligament or ligament complex to the vertical calcified walls. From the attachment region, two cylindrical membranes

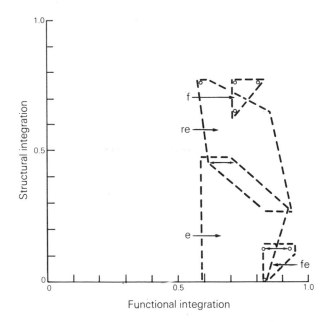

Figure 1.13 Colonial integration of gymnolaemates, arranged by growth form. Structural integration is a compound measure based on zooidal walls, interzooidal connections, and extrazooidal skeleton; functional integration is also a compound measure, based on colony astogeny, development of heteromorphs, and regularity of their position: e, encrusting; f, free-living; fe, flexibly erect; re, rigidly erect. (From Boardman & Cheetham 1973, courtesy of the authors.)

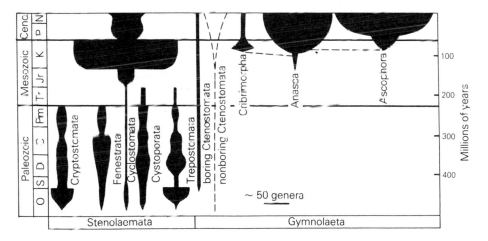

Figure 1.14 Ranges and approximate generic diversity of the higher taxa of stenolaemates (left) and gymnolaemates (right). (Adapted from Boardman & Cheetham 1986, courtesy of Blackwell Scientific.)

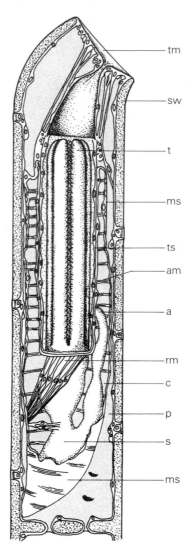

Figure 1.15 Cutaway view of a representative zooid of the Stenolaemata, based on *Crisia*: a, anus; am, annular muscles of membranous sac; c, coelom; ms, membranous sac; p, pseudocoel; rm, retractor muscle; s, stomach; sw, skeletal wall; t, tentacles; tm, terminal membrane; ts, tentacle sheath. (Modified from Nielsen & Pedersen 1979, courtesy of The Royal Swedish Academy of Sciences.)

extend inward: the tentacle sheath which attaches at the base of the lophophore, and the membranous sac, which surrounds the tentacle sheath and the gut. The membranous sac is apparently mesodermal (Nielsen & Pedersen 1979) and therefore encloses a coelom but is surrounded within the zooid by a pseudocoel. The presence of the membranous sac in stenolaemates represents a profound difference in tissue organization between the stenolaemates and gymnolaemates (§ 1.3.2), although it may have originated by a fairly simple division of the body wall (Taylor 1981).

Sequential contraction of annular muscles around the membranous sac in *Crisia, Tubulipora*, and presumably other stenolaemates, squeezes the polypide out as the orifice and the vestibule are being dilated (Nielsen & Pedersen 1979). The base of the lophophore has been observed to move only as far as the approximate level of the skeletal aperture, which has had a profound effect on surface sculpture of colonies (§ 6.3).

Stenolaemate colonies are founded by a single ancestral zooid, the ancestrula, produced by larval metamorphosis. The ancestrula typically begins as a hemispherical disc, flattened against the substratum and with the free, curved surface interrupted by a short circular collar that supports the terminal membrane. The hemispherical disc and collar are defined by an outer cuticle that, except apparently in fenestrates (Gautier 1973), is coated interiorly by calcareous skeleton. Budding of zooids begins with a balloon-like extension of the cuticle. This space is subsequently subdivided by formation of tubular zooidal walls to form a multizooidal budding zone. This is the only form of budding in most stenolaemates. However, in some massive species, new zooids may also be budded adjacent to or interspersed with fully developed zooids, thereby maintaining a constant interzooidal spacing as the surface of the colony increases.

Zooecia in stenolaemates may continue to lengthen throughout the life of a zooid, by addition of calcium carbonate at the terminal aperture. The inner portions are typically thin as a result of relatively rapid growth, and extend subparallel with the local growth direction of the colony. The outer portions of zooecial walls – excepting the isolated orificial collars of some species – are relatively thicker, are extended less rapidly than the inner portions, and grow at high angles to the local colony growth direction. There is generally a curve or sharp bend at the transition from inner to outer parts of zooecia.

In most stenolaemates, elongation of the zooid creates more length than is used by the polypides, which remain near the orificial end. Cycles of polypide degeneration and regeneration produce a series of **brown bodies** that typically accumulate in the inner ends of zooids (Fig. 1.16). Diaphragms in many fossil stenolaemates shorten the living chamber and seal off the excess space in inner ends of zooids. Apparent brown bodies, each sealed behind a diaphragm in a colony of the Devonian trepostome *Trachytoechus*, suggest that diaphragms were associated with degeneration–regeneration cycles in at least some Paleozoic stenolaemates (Boardman 1971). In some stenolaemates, especially those in which zooids remain relatively short, polypides are retracted to inner ends of the zooecia. Brown bodies that result from polypide degeneration in such zooids do not accumulate and are presumably eliminated with the next regenerative stage. This appears to have been the case for the Paleozoic Fenestrata, in which only single brown bodies have been seen in zooecia.

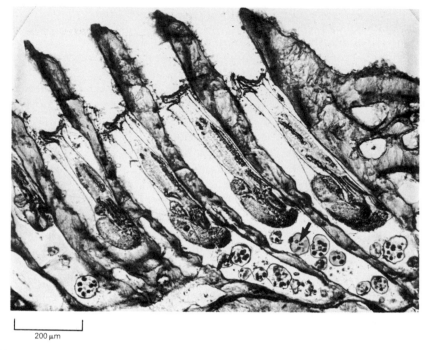

200 µm

Figure 1.16 Undescribed stenolaemate in which the zooids contain several roughly spherical brown bodies (arrowed), which are membrane-encapsulated remnants of former polypides, below the existing polypides; Cook Inlet, Alaska. (Photograph courtesy of R. S. Boardman.)

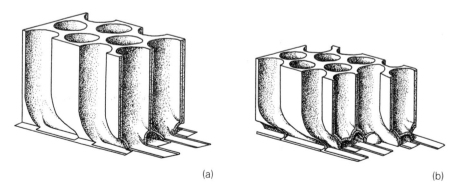

(a)   (b)

Figure 1.17 Idealized diagrams of encrusting and bilaminate stenolaemate colonies in which zooids have distinct encrusting and erect portions. (a) Encrusting portion so short that zooids longitudinally aligned in growth direction (toward left) do not overlap; (b) encrusting portion longer, with overlap of aligned zooids producing complex basal shapes as seen on right. (After Boardman & Utgaard 1966, courtesy of the Society of Economic Paleontologists and Mineralogists.)

The aperture and zooidal chamber have equivalent or nearly equivalent diameters in stenolaemates with tubular zooids. However, among species with sac-shaped zooids, the aperture is smaller than the major part of the chamber and is at the terminus of a relatively narrow tube.

Zooids that originate in marginal and distal budding zones of encrusting and bilaminate stenolaemate colonies have rhombically spaced, parallel, recumbent endozonal portions of varying length (Fig. 1.17, Boardman & Utgaard 1966). The length of the recumbent sections seems to be the sole factor that determines the degree of geometric complexity that exists there.

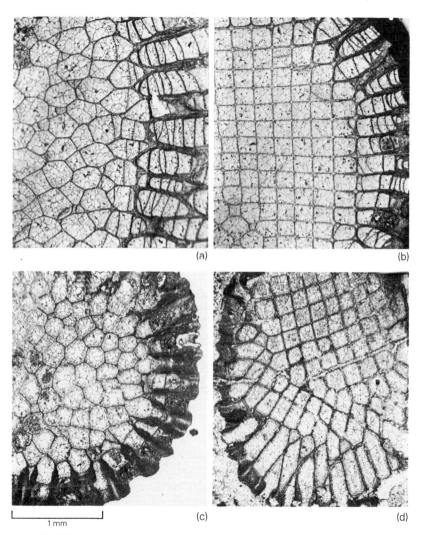

Figure 1.18 Photographs of transverse sections of the trepostomes (a) *Amplexopora*, (b) *Rhombotrypa*, (c) *Tabulipora*, and (d) *Rhombotrypella*. See Figure 1.19 for evolutionary relationships.

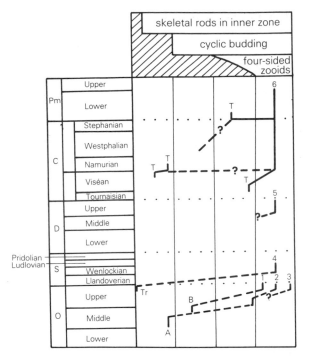

Figure 1.19 Probable morphological stages in evolution of trepostomes with four-sided zooids and their inferred lineages: 1 *Tetratoechus*, 2 *Rhombotrypa*, 3 *Anaphragma*, 4 *Acanthotrypina*, 5 *Eodyscritella*, 6 *Rhombotrypella*, A *Amplexopora*, B *Balticoporella*, Tr *Trematopora*, T *Tabulipora*. (Modified from Boardman & McKinney 1976, courtesy of the Society of Economic Paleontologists and Mineralogists.)

The indentations on the undersides of elongate recumbent zooids are more complex than on zooids with short recumbent portions. These two geometries are widespread in encrusting and bilaminate colonies of Paleozoic and post-Paleozoic age, but their evolutionary and functional significance within the class are unknown. It is possible that they simply reflect growth rate, with more elongate endozonal portions resulting from faster growth rate.

Several repetitively evolved zooidal geometries and distributional patterns also characterize endozones in radially symmetrical branches both within and between the stenolaemate orders (McKinney 1975, 1977a). The most extensively studied of these endozonal patterns has a reticulate arrangement of four-sided zooids (Fig. 1.18). A basic attribute of the pattern is the regular doubling in number of zooids at convex zones in which one new zooid forms at each corner between four previously established, neighboring zooids (Boardman 1968, Boardman & McKinney 1976). The six known trepostome genera characterized by four-sided zooids belong to six different lineages as indicated by a complex of exozonal characters, and reasonable ancestral grades are known for three of them (Fig. 1.19). The most common, and apparently most primitive, types of extended budding zones in radially symmetrical branches generate new zooids at unpredictable points (McKinney 1977a). As in encrusting and bilaminate forms, evolutionary and possible functional significance of the various endozonal geometries are unknown.

The cuticle that surrounds a stenolaemate colony is attached to the skeleton across the underside of the colony base. Above the base in many living colonies the cuticle is unattached to the mineralized skeleton except where it is folded into zooids and attached radially by ligaments *within* zooidal chambers. These are the **free-walled** stenolaemates. In contrast, most post-Paleozoic and very few Paleozoic forms have an additional calcified strip of frontal wall. Where such calcified frontal wall occurs, and also where basal or lateral walls are in direct, continuous contact with the cuticle above the colony base, there is no outer fluid-filled space. These are the **fixed-walled** stenolaemates. The different types of wall organization in stenolaemates determines or limits the paths of nutrient flow within colonies, as discussed in § 1.2.1. For more information on stenolaemate organization, including complexities not covered here, see Borg (1926, 1933), Nielsen & Pedersen (1979), and Boardman (1983).

Five orders constitute the class Stenolaemata. They are the Trepostomata, Cystoporata, Cryptostomata, Fenestrata, and Cycloutomata. Some stenolaemate taxonomists do not recognize Fenestrata separately from Cryptostomata, while others divide Cryptostomata as used here into Cryptostomata (s.s.) and Rhabdomesonata. Soft parts of zooids are thought, with some evidence even for apparently extinct orders, to be generally similar within all orders of the class (McKinney 1969, Boardman 1971, 1983, Utgaard 1973, Boardman & McKinney 1976, 1985, McKinney & Boardman 1985). The orders are distinguished on the basis of zooidal geometry, colony geometry, skeletal structure, and types of polymorphs and extrazooidal structures (Table 1.1, Fig. 1.20). The post-Paleozoic cyclostomes are possibly a polyphyletic hodge-podge of survivors of the trepostomes, cystoporates, Paleozoic cyclostomes and, perhaps, cryptostomes (Boardman 1981, 1984).

Table 1.1 Characteristics of the marine Bryozoa (ranges and characteristics after Boardman *et al.* 1983, Boardman 1984).

Class **Stenolaemata** – Zooids are elongate cylindrical and continue to lengthen through ontogeny, with long axis at angle to local colony growth direction; basal and vertical walls rigidly calcified; interzooidal communication through space between outer ends of vertical walls in most, but through pores in vertical walls in some; membranous sac encloses polypide and is deformed to protrude lophophore through orifice at outer end of skeletal tube. Lower Ordovician–Recent.

Order **Trepostomata** – colonies encrusting or erect; elongate autozooids generally containing basal diaphragms and commonly other lateral structures; kenozooids common, extrazooidal skeleton in some; skeletons without communication pores, typically laminated. Lower Ordovician–Upper Triassic or ?Recent.

Table 1.1   (cont.)

Order **Cystoporata** – colonies encrusting or erect; autozooids short and without, or long and with, basal diaphragms; most autozooids with thickened strip (**lunarium**) of different microstructure along one side in thick-walled outer zone; skeletal structure laminated, granular or granular–prismatic, some with communication pores; kenozooids in some, gonozooids uncommon; vesicular extrazooidal skeleton common and abundant. Lower Ordovician–Upper Permian or ?Cretaceous.

Order **Cryptostomata** – colonies erect arborescent, or bilaminate sheets; generally short autozooids, some with basal diaphragms or **hemisepta** (incomplete lateral partitions); skeletons without communication pores, typically laminated; kenozooids or extrazooidal skeleton may be present. Lower Ordovician–Upper Permian or ?Cretaceous.

Order **Fenestrata** – colonies erect, composed of narrow unilaminate branches; short autozooids commonly with hemisepta, but basal diaphragms in very few; primary zooidal skeleton without communication pores, typically granular; heterozooids (gonozooids, nanozooids, and various others) in some; extrazooidal skeleton extensive, several times autozooidal skeleton in volume, consisting of laminae penetrated by small granular rods. Lower Ordovician–Upper Permian or ?Triassic.

Order **Cyclostomata** – colonies encrusting or erect; autozooids commonly long, some with basal diaphragms and other structures; skeletal structure typically laminated, many with communication pores; gonozooids common, kenozooids or nanozooids in some; extra-zooidal skeleton may be present. Lower Ordovician–Recent.

Class **Gymnolaemata** – zooids generally box- or sac-shapes to short cylinders, with long axis roughly parallel to local colony growth direction; zooidal size is fixed early in ontogeny; zooidal body walls entirely organic to rigidly calcified; interzooidal communication by funicular network through tissue-plugged pores in vertical walls; vertical or frontal walls deformed to protrude lophophore. Upper Ordovician–Recent.

Order **Ctenostomata** – zooidal walls membranous or gelatinous; orifice terminal and closed by pleated collar in most; heterozooids absent or only kenozooids. Upper Ordovician–Recent.

Order **Cheilostomata** – zooidal walls calcified, flexible or rigid; orifice frontal and closed by proximally hinged operculum; heterozooids usually present, commonly diverse; suborders based on frontal calcification and mechanism of lophophore protrusion (§ 1.3.2.). Upper Jurassic–Recent.

1.3.2   Gymnolaemata

Autozooids of the gymnolaemates range from box- or sac-shapes to short cylinders (see Figs. 1.1, 1.2). Their principal axis generally coincides with the local growth direction of the colony. The direction of growth is defined as distal, and that from which it came is **proximal**. The **frontal** direction is defined by the surface that contains the orifice; the opposite is **basal**. Walls between zooids (**vertical walls**) are roughly uniform in construction from basal to frontal sides, and thickness either does not vary or grades uniformly from one side to the other. Zooids are interconnected through cell-plugged pores in the vertical walls. Growth of the walls is completed early in ontogeny, establishing the dimensions of the zooidal living chamber before a polypide is budded. Throughout ontogeny the polypide, when present, is retractable to an approximately constant position within the coelom-filled

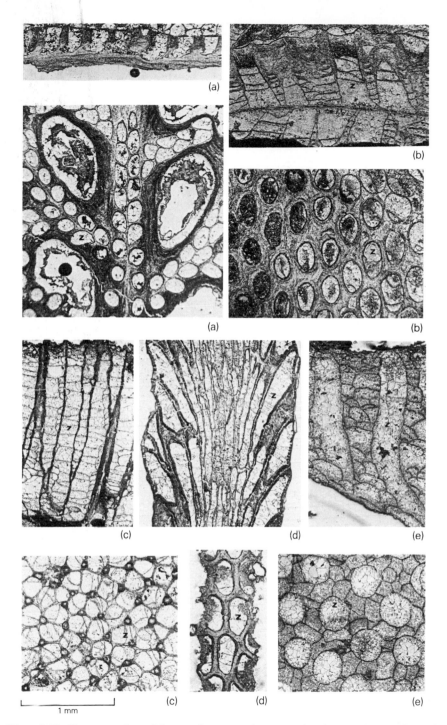

Figure 1.20 Representatives of the stenolaemate orders as seen in paired longitudinal (above) and tangential (below) sections. The entire length of the tubular zooids is seen in longitudinal section, and cross sections of zooids are seen in tangential sections. A single zooid in each is marked by a "z". (a) *Septopora*, Order Fenestrata; (b) *Arthrophragma*, Order Cryptostomata; (c) *Mesotrypa*, Order Trepostomata; (d) *Collapora*, Order Cyclostomata; (e) *Fistulipora*, Order Cystoporata.

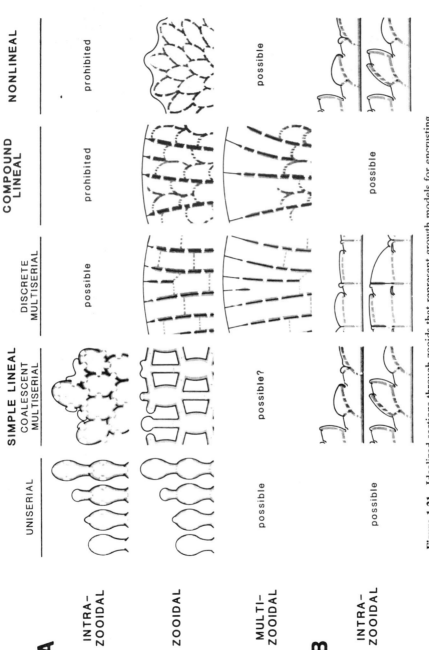

Figure 1.21 Idealized sections through zooids that represent growth models for encrusting cheilostomes, based on patterns of zooidal budding. The morphogenetic states illustrated may occur singly or combined within species or colonies. Rows represent budding processes, and columns represent budding geometries. Single-layered states are illustrated as tangential sections, and multilayered states are illustrated as longitudinal sections. Interior walls are shown with skeleton only, whereas cuticle (black lines) and skeleton (stippled) of exterior walls

...(After .... Lidgard 1985; courtesy of the Palaeontological Association.)

chamber. Other generally widespread characters of the class include polymorphism, larval brooding, and budding of zooids in lineal series (Ryland 1970, Cheetham & Cook 1983).

The gymnolaemates are divided into two orders, the Ctenostomata and the Cheilostomata (Table 1.1). Ctenostome zooids are cylindrical, sac-shaped, or flat. Walls are not calcified but are membranous or gelatinous, with a terminal orifice closed by a pleated collar. Heterozooids include stolons, which are a form of kenozooid, but these are present only in one suborder. Most living ctenostomes grow as miniature trees or as vines creeping across the substratum (see Fig. 1.12), but there are also massive encrusting forms, such as many species of *Alcyonidium*. Some ctenostomes provide themselves with a skeleton by boring into shells of other organisms, most typically brachiopods in the Paleozoic and molluscs in more recent times.

The fossil record of ctenostomes begins at least as early as Late Ordovician and consists primarily of excavations made by stoloniferous borers (Pohowsky 1978). Nonboring ctenostomes are known from relatively few instances in Jurassic and younger strata, where their shape was preserved on the undersurface of some other organism that grew over them (§ 7.6, Voigt 1979).

Cheilostome zooids (see Fig. 1.2) are generally flattened box- or sac-shaped, although a few form short cylinders. Walls are calcified. In most species, the frontal and basal walls have greater areas than do the vertical walls. The orifice is closed by an operculum that is hinged near the proximal side and is subparallel with the zooidal growth direction when closed. Heterozooids are generally abundant, especially avicularia and hetero-zooidal brooding structures.

Three general types of zooidal budding are known in cheilostomes (Fig. 1.21, Lidgard 1985a, b). **Intrazooidal budding** occurs where a space originally included within the parent zooid is transformed into the proximal part of a new zooid. This is accomplished by swelling of the adjacent outer cuticular and cellular layers of the parent to define the space of the new zooid's walls. The space that changes from parental to descendant zooid may be a lateral pore chamber or the coelomic space between frontal cuticle and a calcified shelf-like wall below, so that intrazooidal budding may extend the colony horizontally or vertically. **Zooidal budding** occurs where cuticular and cellular layers of exterior wall of the parental zooid(s) grow outward to extend the body cavity. A perforated interior wall grows to partition the new zooid from the parent, and calcification of the body walls of the new zooid follows soon after budding. **Multizooidal budding** is similar to zooidal budding except that the cuticular and cellular layers of the exterior wall include uninterrupted space beyond the last-formed zooid(s) that is eventually partitioned into two or more zooids by interior walls. The separation of multizooidal and zooidal budding is arbitrary, because there are intermediate forms. The concept of multizooidal budding is useful, however, because it occurs in relatively few taxa, it is related to growth rate, and it is of considerable ecological importance (Ch. 7).

In addition to budding processes, geometrical relationships between pre-existing and newly developed zooids show important variation (Fig. 1.21). **Simple lineal arrangements** occur where zooids are bounded laterally by exterior wall consisting of cuticle and cellular layers, and usually an intermediate calcareous layer. There are three basic variations of simple lineal arrangements. In uniserial growth, each lineal series is spatially isolated from others within the colony. This is apparently the ancestral state for the two remaining variations. In coalescent, multiserial growth, buds from two or more parental zooids touch and fuse into a single developing zooid, resulting in a matrix of parent–descendant relations within a two- or three-dimensionally continuous colony. Lastly, in discrete, multiserial lineal arrangements, different contiguous lineal series each retain straight-line parent–descendant relations, and communication between adjacent series occurs through interserial communication pores developed by resorption of the intervening cuticular walls. **Compound lineal arrangements** contain two or more rows of zooids separated from one another by interior walls within advancing multiserial units that are bounded by exterior walls. **"Nonlineal" arrangements** occur where zooids develop entirely within a confluent budding zone so that each is separated from its neighbors only by interior walls. Although the developing zooids in "nonlineal" arrangements may have direct contact with only two or three pre-existing zooids, the space in which zooids develop has direct contact with, and is presumed to receive nutrients from, numerous mature zooids.

Primitively, cheilostome zooids are surrounded entirely by cuticle-bounded exterior walls, except for the small communication pores that locally breach the walls between adjacent zooids. The earliest known genus, the Late Jurassic *Pyriporopsis* (Fig. 1.22), is uniserial and has proximally-tapered ovoid or pyriform zooids that are budded intrazooidally. Budding of new zooids from more than one distal and disto-lateral pore chamber generates multiserial, sheet-like colonies composed of exterior-walled zooids from coalescence of buds from adjacent series. This first occurred very early in the history of cheilostomes as seen in *Wilbertopora* from the Early Cretaceous (Albian) of Texas, in which proximal parts of the colony may be uniserial and more distal portions are multiserial (Fig. 1.23).

The most common mode of budding in cheilostomes produces lineal series bounded laterally by exterior walls (Boardman & Cheetham 1969, 1973, Cheetham & Cook 1983). In multiserial colonies that are organized in lineal series, each series swells distally, more or less growing at the same rate as, and in step with, advancement of the adjacent series so that a smooth colony margin is maintained. This growth pattern first evolved late in the Early Cretaceous (Boardman & Cheetham 1973, Lidgard 1985b).

Two variables determine whether growth in simple lineal series is zooidal or multizooidal: rate of advancement of the distally expanding ends of the

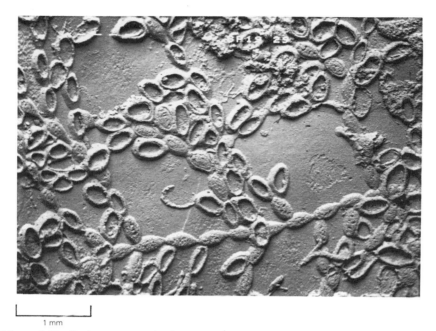

└─────────────┘
1 mm

Figure 1.22 *Pyriporopsis portlandensis*, the oldest known cheilostome, Upper Jurassic (Portlandian), Whitchurch, Bucks., England. The colony form and zooidal simplicity of *Pyriporopsis* suggest an origin from runner-like ctenostome ancestors (Pohowsky 1973, Banta 1975, Cheetham & Cook 1983). (Photograph courtesy of P. D. Taylor.)

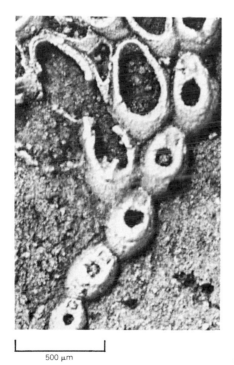

└─────────────┘
500 μm

Figure 1.23 *Wilbertopora mutabilis*, an early, multiserial anascan cheilostome with earliest stages of colony astogeny resembling ancestral uniserial state; Kiamichi Formation (Albian, Cretaceous), Patton, Texas. (Photograph courtesy of A. H. Cheetham.)

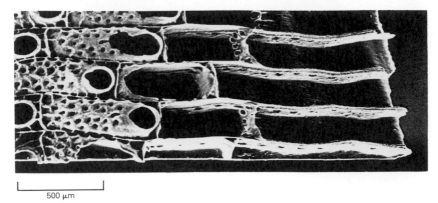

500 μm

Figure 1.24 Discrete multiserial, multizooidal budding geometry in *Schizoporella floridana*, Pescaderabaai, Curacao. (From Lidgard 1985b, courtesy of the Palaeontological Association.)

lineal series, and rate of growth of pore-bearing transverse interior walls (Lutaud 1961, 1983). If the latter lags behind the former, multizooidal budding occurs, with a graded series of partially differentiated zooids lagging behind the rapidly expanding tip or "giant bud" (Fig. 1.24). Giant buds in the anascan *Membranipora membranacea* at the end of each lineal series may be several millimeters long, and the daily progression of the growing edge is roughly twice the length of the bud. Zooid length, generally between 0.8 and 1.2 mm, correlates with growth rate, and a new zooid is partitioned off at the proximal end of the giant bud every 4–6 hours.

Major distinctions between higher taxa of bryozoans, as well as much evolutionary history, is embodied in solutions devised for protrusion of tentacles from calcified zooids (Cheetham 1971, Taylor 1981). Tentacle protrusion in all bryozoans is accomplished by reducing the volume of the living chamber, which causes the lophophore end of the polypide to be squeezed out of the orifice. In cases where the tentacle sheath is relatively long, the gut, too, may clear the orifice, encased within the everted sheath.

Tentacle protrusion in ctenostomes is straightforward, since all parts of the wall are potentially deformable. Parietal muscles, particularly transverse muscles, attach to the inner surface of the body wall. Their contraction reduces the diameter of the cylindrical zooid, which causes part of the body volume to be pushed toward the orifice, everting the tentacle sheath so that the tentacles protrude and can open to the feeding position. Retraction of polypides occurs when the retractor muscle inserted at the base of the tentacles contracts powerfully, with simultaneous relaxation of parietal muscles.

In *Pyriporopsis* and other cheilostomes with an uncalcified frontal wall, parietal muscles connect the frontal wall and the lateral portions of the vertical walls (Fig. 1.25a). Contraction of the parietal muscles depresses the frontal wall, which displaces body fluid, forcing protrusion of the lophophore end of the polypide out of the orifice as the operculum rotates

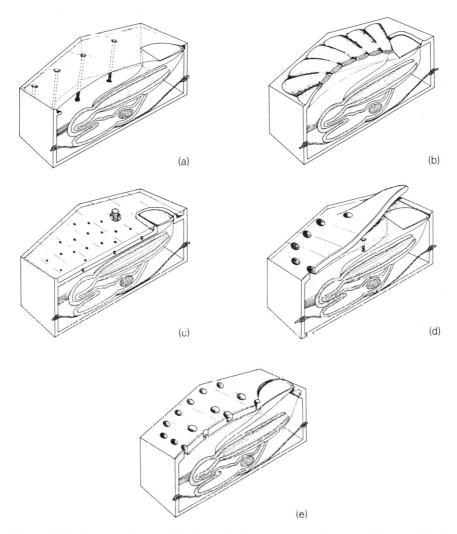

Figure 1.25 Representative zooids of the Cheilostomata. (a) Anascan with noncalcified frontal wall and polypide unprotected by shield. (b) Anascan (Cribrimorpha) with shield of fused spines arched above the uncalcified frontal wall. (c) Anascan with perforated shield below the uncalcified frontal wall. (d) "Umbonuloid" ascophoran with a calcified shield arched above the uncalcified frontal wall. (e) Ascophoran with calcified frontal wall and inflatable ascus opening through a well-defined pore covered by the proximal portion of the operculum.

open. The process is more complicated for most cheilostomes with calcified frontal walls. Most of these possess an inflatable sac-shaped cuticular balloon, the **ascus**, which is infolded in various ways from the frontal wall (Fig. 1.25d, e). Contraction of parietal muscles connecting the ascus and vertical body walls distends the sac, which displaces coelomic fluid and pushes out the polypide.

Cheilostomes with an ascus are polyphyletic, but are classified together for convenience in the suborder Ascophora. In contrast, cheilostomes without an ascus comprise the suborder Anasca. These appear earlier in the fossil record, and are clearly more primitive.

Calcification of the frontal region has occurred in several evolutionary lineages of cheilostomes. Reinforcement of the broad frontal region affords increased protection to retracted polypides from predators or physical injury, and, for erect forms, enhances the strength of zooids that bear stresses from more distal parts of the colonies. However, a completely calcified set of walls, including the frontal, could not allow the volume reduction needed to push out the lophophores to the feeding position. Different types of **frontal shields** (Fig. 1.25) have evolved that allow both functions.

Some cheilostomes have hollow, coelom-filled marginal spines that arch over the flexible frontal membrane. In extreme cases, these spines extend from the margins to the midplane above the frontal wall to touch or fuse with those from the opposite margin. These spines may also have lateral projections that touch and fuse with those of the neighbouring spines. Thus, the frontal surface is covered almost entirely, with regularly spaced holes between spines to allow water to flow into and out of the space above the flexible frontal wall (Fig. 1.25b). Many taxonomists recognize a distinct suborder, the Cribrimorpha, for such species with fused spines over the frontal wall. These bryozoans grow almost exclusively as single-layered, encrusting colonies.

Other anascan cheilostomes have a perforated shelf formed below the uncalcified frontal wall and above the position of the retracted polypide (Fig. 1.25c). Parietal muscles attach to walls below the shelf and connect to the frontal wall by ligaments through pores in the shelf. The shelf therefore protects the polypide but leaves the flexible frontal wall exposed.

Most cheilostomes with an extensively calcified, nondepressible frontal shield have an ascus. In some of these ascophorans, the ascus is protected on the exterior by a calcified shelf that arches over the frontal wall from the zooidal margin (Fig. 1.25d) and has a thin, cuticle-enclosed coelomic space over it. The fold extends all round the lateral and proximal sides of the zooid, and water flows into and out of the space between it and the frontal wall at its open, elevated margin just proximal to the orifice. In other ascophorans, the ascus opens through a small, isolated pore proximal to the orifice, or the operculum operates as an excentric butterfly valve, with the smaller portion covering the opening of the ascus, proximal to the fulcrum points (Fig. 1.25e). As the larger part of the operculum over the orifice rotates up for passage of the lophophore, the smaller part over the ascus rotates down, allowing water to flow in.

There is an overwhelming diversity of patterns of development of frontal shields among cheilostomes, such that the spatial relations between wall, skeleton, coelom, and cuticle may be completely different from case to case.

Different types of shields have arisen repeatedly, even within a single clade (Banta & Wass 1979). Formation of such protective frontal shields has been a dominant theme in cheilostome zooidal evolution, and is important for the potential to bud frontally and to strengthen colonies sufficiently for robust erect growth.

2 Species relationships and evolution

The functional, ecological, and evolutionary patterns described in this book are based on analyses of many higher taxa without concern for taxonomy at the species level. Understanding how such patterns come about, however, requires detailed knowledge of species relationships and patterns of evolution of species within well-defined individual clades.

Species of calcified bryozoans have been defined predominantly on the basis of qualitative description and linear measures of skeletal morphology of zooids, characters whose functional importance is at best poorly understood. Subtle features of colony form are seldom noted but are potentially useful in species characterization. Nonskeletal attributes have been virtually ignored for living species. The few attempts to resolve questions about species recognition using studies of inheritance and electrophoresis suggest that species with greater skeletal complexity may be, in general, more realistically defined than those with simpler skeletons.

Studies of species evolution and inferences of species lineages are rare for bryozoans. The most detailed and comprehensive study traces the history of *Metrarabdotos*, which reveals clearly marked rates of change and evolutionary patterns. The general lack of information on species evolution is matched by little agreement about relationships among the higher level taxa of bryozoans in current classifications.

2.1 Morphological characterization of species

The overwhelming majority of living bryozoan species, as well as all fossil species, are defined solely on morphology, as is true for most organisms. In this section we describe the typical approaches to morphological characterization of species and present some preliminary evaluations of their ability to discriminate genetic species.

2.1.1 Commonly used characters

Most descriptions of living bryozoans with calcified skeletons do not differ substantially from descriptions of fossil species in the same class. This is because the majority of taxonomists who work on living bryozoans have depended on skeletal characters for species recognition, to the virtual exclusion of soft organs and other attributes. The dominant focus has been on morphology of zooids. In addition to presence or absence of structures, a variety of linear measures of size, shape, skeletal construction, and types of heterozooids have been used for both living and fossil species (see, for example, Cheetham 1966, Cuffey 1967, Walter 1969, Soule & Soule 1973, Harmelin 1976, Ryland & Hayward 1977, Hayward & Ryland 1979, 1985, Hayward & Cook 1979, 1983, Vavra 1983, Karklins 1986).

Living stenolaemates are known almost exclusively from external skeletal morphology. Arrangement and diameters of autozooidal apertures; placement and sizes of heterozooids such as kenozooids and nanozooids, and of brood chambers and their orifices; and growth form and robustness have been considered of primary importance as criteria for species characterization. Internal skeletal features have been considered occasionally (Borg 1933, Brood 1976, Harmelin 1976, Farmer 1979), and polypide and other soft structures have been added in a few studies (e.g. Borg 1933, Harmelin 1976, Boardman & McKinney 1985). Studies of post-Paleozoic stenolaemates have similarly emphasized external skeletal morphology with few exceptions (Walter 1969, Brood 1972, Hinds 1975, Tillier 1975, Nye 1976). In contrast, most Paleozoic stenolaemates (Trepostomata, Cystoporata, Cryptostomata) have been studied and characterized predominantly from oriented thin sections and acetate peels. These different considerations of morphology in taxonomy seem to be more a matter of historical precedent than a logical incorporation of all available characters.

Cheilostomes, excepting the simpler membranimorphs, are skeletally complex, with several potential species-level features such as details of colony form, zooidal size, various conditions and details of the frontal shield, shape of primary and secondary orifices, and polymorphic features, especially ovicells and avicularia (Larwood 1962, Cheetham 1968, Hayward & Cook 1979, 1983, Cheetham et al. 1981, Vavra 1983). Number, sizes, shapes, placement, and orientation of avicularia can be particularly complex and useful, as in distinguishing the various species of *Metrarabdotos* (Fig. 2.1).

Growth form and colony size have been considered with disfavor as species characteristics because of the possibility of profound environmental influence on them. However, there are indications across a broad range of bryozoan taxa that these may be useful occasionally. For example, among trepostomes in the Devonian Hamilton Group of New York, 20 had stable growth forms (either encrusting or arborescent with cylindrical branches) whereas six were both encrusting and erect (Boardman 1960). Within the fauna, details of form differ between species, e.g., *Atactotoechus parallelus* has more robust branches with a smaller proportion of the branch width occupied by the thin-walled interior zone than *A. acritus*. These different measures and proportions most likely result from different average branch diameters and degree of taper at growing tips of branches.

The "rules" by which species of Cenozoic and living adeoniform cheilostomes grow yield colony morphologies that are species specific (§ 8.3.2, Cheetham et al. 1980, 1981). Although there is considerable variance

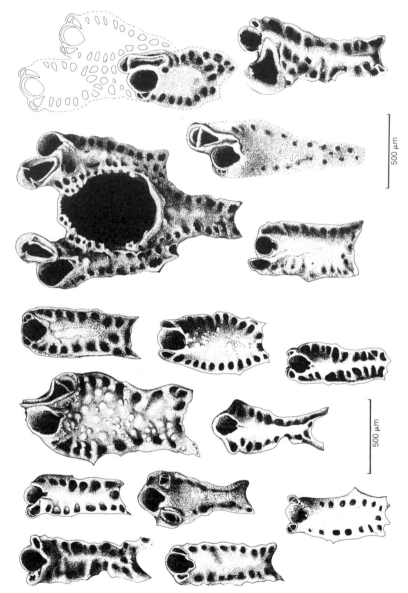

Figure 2.1 Examples of differences in ordinary (left) and special (right) avicularia among species and subspecies of *Metrarabdotos*. (From Cheetham 1968, courtesy of the Smithsonian Institution Press.)

in link length and angle of branching within species, differences between species are consistent and detectable. Branching parameters set both shape and, through increased branch interference, maximum possible sizes for adeoniform colonies (Cheetham & Hayek 1983). There is also a maximum characteristic size for many discoidal free-living species in which colonies are concentrically zoned, with an abruptly downcurved perimeter of unique heterozooids (§ 9.1.2, Chimonides & Cook 1981). In such cases, maximum colony size may be a good taxonomic character.

Persistent but subtle differences in colony form may also exist for encrusting species that seem to grow to shapes determined by the local environment, although they normally go undetected for lack of attention. Their existence even between sibling or closely related species is indicated by the "morphotypes" of *Parasmittina nitida* having different propensities for frontal budding in identical controlled environments (Humphries 1975).

2.1.2 Missed opportunities

Even now, with the benefits of statistical description and comparison in taxonomy well known for over 25 years, and computers accessible by virtually everyone, descriptions of bryozoans may involve no more than the most perfunctory measurements. Descriptions of stenolaemates commonly contain the basic univariate descriptions and have occasionally invoked *t*-tests for comparison of means, but with rare exceptions (e.g., Cheetham 1966, 1968) descriptions of cheilostomes are less informative about measurements. Scatter diagrams showing plots of relative change and degree of correlation are even less common.

Species descriptions are accompanied by almost no mention of why particular attributes have been singled out for study while others are omitted. The lack of descriptive consistency and insight into authors' taxonomic concepts has severely hampered studies of bryozoan evolution at low taxonomic levels. In part the situation is understandable, because so much description has been for little-known faunas such as those living off the south-east coast of Africa (Hayward & Cook 1979, 1983) or preserved in the Upper Paleozoic of the American west (Karklins 1986) or Mongolia (Goryunova & Morozova 1979), and experienced taxonomists have felt it important to make the faunas known rather than to dwell on detailed quantitative description and discrimination of species. However, the shallow-water faunas of the North Atlantic and the Pacific coast of North America are relatively well known; study and description of them as well as monographs on particular genera would benefit from more systematic quantitative description and analysis, and detailed knowledge of embryos, life histories, behaviors, habitats, and molecular characterizations. A systematically accumulating store of these types of information is needed for continued reassessment of the value of skeletal morphology in recognition of living and fossil bryozoan species (§ 2.2).

Several straightforward multivariate techniques are available and have been used successfully, but infrequently, to evaluate species groupings of populations of colonial animals (e.g., Brood 1972, Moyano 1972, Cheetham 1975, Cheetham & Lorenz 1976, Cook 1977b, Foster 1984, 1985, Pachut & Anstey 1984). When properly applied (Brande & Bretsky 1982) to bryozoans and corals, techniques such as principal components analysis, canonical variate analysis, and multivariate analysis of variance can group specimens and species with less subjectivity, reduce measures and counts to a smaller number of variates that minimize redundancy, and identify the relative contribution to overall variation from various sources. Foster (1984, 1985) gives a clear statement of the sequence of use and the contribution of several of the techniques to recognition of species in Neogene hermatypic corals and hypotheses about their phenotypic responses to the environment.

2.2   Is morphology adequate?

Modern calcified bryozoans that are illustrated in taxonomic papers have been soaked in sodium hypochlorite in order to dissolve the highly reflective outer cuticle, a process that also dissolves all other noncalcified tissues. As a result of this treatment, and most taxonomic work having been done on preserved samples, tentacle number and color of colonies are described for a very small proportion of species. Polypide measures, behavioral characteristics, detailed ecological and reproductive data, and other biological attributes have been reported for relatively few and are seldom taken into account in systematics, although they are known to be commonly species-specific. Therefore, in addition to the absence of information on living taxa, an extensive and systematic basis for paleobiological inferences is missing as well.

The skeletal complexity of cheilostomes commonly yields over 20 characters that have been considered useful in distinguishing congeneric species, in contrast with half that number, or even less, for stenolaemates and ctenostomes, as can be seen clearly by comparing descriptions in various taxonomic monographs published during the past 20 years. This calls into question how well biological species can be recognized from skeletal data and whether cheilostome and stenolaemate species can be discriminated with equal success on the basis of skeletal information alone.

2.2.1   Reliability of skeletal morphology based on breeding

Controlled interbreeding experiments have yet to be carried out for bryozoan species, in large part because outbreeding is difficult to assess and control in this largely hermaphroditic phylum. However, laboratory studies of maternal inheritance in the cheilostome *Parasmittina nitida* suggest that two previously differentiated but sympatric "morphotypes" are genetically

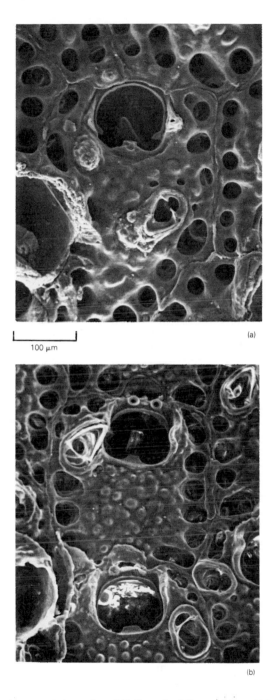

100 μm

(a)

(b)

Figure 2.2 Morphotypes A (upper) and B (lower) of *Parasmittina nitida* (from Morehead City, North Carolina), showing differences in number, shape and placement of avicularia, and in shape and sculpture of the orifice.

isolated (Maturo 1973, Humphries 1975). The two forms are skeletally distinguished on morphology of zooidal orifices; shape, sizes, number and placement of adventitious avicularia (Fig. 2.2); and morphology of the ovicell. (In addition, Humphries found numerous conspicuous differences in larvae of the two morphotypes.) In the original study, 50 maternal colonies (29 of one morphology and 21 of the other) were collected in the field and brought into the laboratory. Altogether they released 2053 embryos; 438 of these offspring survived to produce colonies with recognizable morphology. Each of these developed the same morphology as its maternal colony, indicating genetic incompatibility of the two morphotypes and therefore existence of two distinct species. The second experiment likewise had 100% correspondence of morphotype of offspring to maternal colonies, into the third generation. This pattern was evident at two sites on the mid-Atlantic and Gulf of Mexico coasts of the U.S.A.

2.2.2 Implications of electrophoresis

Gel electrophoresis can be used to estimate numbers of alleles per gene locus by determination of different states of specific enzymes. Variations in electrical charge, and to a much lesser extent weight, determine rate of movement of an enzyme along a gel surface whose pH has been carefully adjusted so that enzymes will migrate across it. After a set migration interval, stain is applied for the specific enzyme being studied so that its positions can be determined. It is assumed that each electrical variant of the specific enzyme reflects a different allele at the controlling gene locus. Electrophoretic studies of two bryozoan taxa in particular, the cheilostome *Schizoporella errata* and the ctenostome *Alcyonidium*, have yielded important information on speciation, the relationship of genetics to morphology, and the rapidity of genetic accommodation to environmental gradients in bryozoans with low dispersal potentials because of short-lived, nonplanktonic larvae.

Along the east coast of North America, shallow subtidal *Schizoporella errata* exhibit temperature-related clines in allele frequency of the enzymes leucine amino peptidase and glutamate oxalate transaminase (Gooch & Schopf 1970, 1971, Schopf & Gooch 1971, Schopf 1974, Schopf & Dutton 1976). The clines are strongly developed along Cape Cod (Figs. 2.3, 2.4) and were maintained over a period of several years. They correlate with maximum summer temperature, which along Cape Cod is 5 °C higher at the south-west than at the north-east. There is also a cline in length of adventitious avicularia that corresponds with the cline in temperature and distance along the Cape, with larger avicularia in relatively cooler water. Gene frequencies and avicularian length change over very short distances. A 1 °C difference in maximum summer temperature, or a 10–13 km distance along shore, is correlated with recognizable differences in allele frequencies and in avicularium length. In contrast, there is no cline in zooid width, orifice

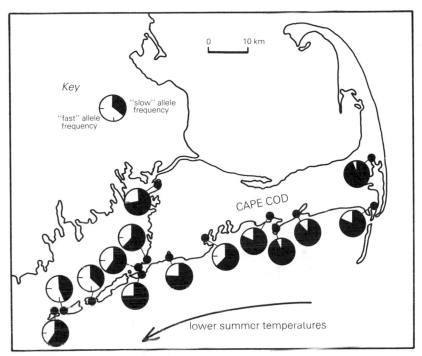

Figure 2.3 Frequency distributions of "slow" (black wedges) and "fast" (white wedges) alleles of leucine amino peptidase in colonies of *Schizoporella errata* along the shore of Cape Cod of 1972. (From Schopf 1974, courtesy of the Marine Biological Laboratory, Woods Hole.)

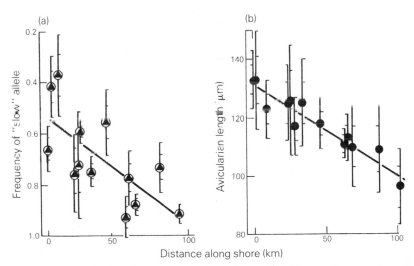

Figure 2.4 Correspondence of genotype and morphotype in *Schizoporella errata* along Cape Cod: (a) frequency of "slow" allele of leucine amino peptidase and (b) mean of avicularium length. Vertical bars represent two standard deviations. (From Schopf & Dutton 1976, courtesy of the Paleontological Society.)

width, or orifice length, suggesting that length of adventitious avicularia varies independently of overall size of zooids and that overall zooid size is not temperature related.

Despite the limited and preliminary state of evidence, these results suggest that small, genetically based morphological differences should be detectable in well-documented lineages of fossil cheilostomes. Although the morphology closely tracks temperature or some temperature-related phenomenon, it cannot be dismissed as environmentally induced variation superimposed on genetically indistinguishable populations (but see § 6.1).

In contrast to the close link between morphology and genetics in *S. errata* and the apparent lack of cryptic species within it, the morphologically simple carnose ctenostome *Alcyonidium* has a distinct propensity for cryptic speciation. Its colonies grow as encrusting sheets, lumpy mounds, lobes, or branched structures composed of polygonal, crowded zooids with gelatinous walls. Taxonomists typically list 5–11 (mean = 7.8) characters by which species within the genus may be distinguished (Osburn 1952, Maturo 1957, Kluge 1962, Ryland 1963, Winston 1982, Hayward 1985). Among British *Alcyonidium* traditionally assigned to *A. gelatinosum*, *A. mytilis*, *A. polyoum*, and *A. hirsutum*, electrophoretic analysis indicates 12 genetic species (Thorpe *et al.* 1978a, b, c, Thorpe & Tyland 1979). In some cases morphological differences can be recognized for genetic species within what had formerly been considered a morphologically variable species, and ecological distinctions can be made for others that are morphologically indistinguishable. However, some of the genetic species are sympatric and indistinguishable morphologically on the basis of the characters examined so far. Commonly, among these latter species, there is no genetic similarity at all for the loci tested, and therefore they are almost certainly not even sibling species but are distantly related within the genus.

2.3.2   Undetected stenolaemate species are probably common

The results of the breeding experiments on *Parasmittina nitida*, and the correspondence of morphology to allele frequency in *Schizoporella errata*, suggest that there is validity in taxonomic "splitting" of bryozoans based on small but persistent differences in skeletal morphology. Many cheilostomes have been divided into sympatric subspecies or varieties that probably represent distinct species.

The small number of characters used by taxonomists to distinguish species of *Alcyonidium* approximates the number typically noted for species recognition within genera of fossil and living stenolaemates. An average of three genetic species per morphologic species in *Alcyonidium* suggests that cryptic species may be rampant in the similarly morphologically simple stenolaemates. The problem is very similar to that which plagued the study of freshwater triculine mesogastropods of the Mekong River, which has undergone a spectacular endemic radiation of over 90 species within the past 12 million years (Davis 1979, 1981). These snails have very simple shells, and convergence of shell morphology of distantly related species within the subfamily is extreme. Only four of the 28 characters used for species identification are skeletal. Although the large number of isolated environments probably contributed to the extraordinary radiation of the triculines in the Mekong River, it is sobering to contemplate variations in soft-part morphology of stenolaemates, such as presence or absence of a gizzard and relative elongation of polypides, that have not been reflected in the skeleton (Boardman & McKinney 1985). Although soft-part morphology, and especially molecular techniques, may contribute greatly to taxonomy of living stenolaemates, it seems inevitable that taxonomists working with fossil stenolaemates will have to accept that extinct "species" within the group are likely to include more than a single genetic species.

All the results summarized in this section (2.2) must be considered preliminary because there are so few. There is a profound need for extensive and detailed morphological studies with well-reasoned statistical analyses and gel electrophoretic studies (and other more powerful molecular techniques) on the same animals. Only then can we evaluate what well-defined fossil "species" mean. Because of the contrast in morphological complexity, transfer of confidence level from taxonomic studies on cheilostomes to inferences for stenolaemates would be misleading. Such comprehensive studies must be carried out on living cyclostomes if any sense is to be made of Paleozoic stenolaemate species, and especially of the tangled and commonly overextended post-Paleozoic cyclostome species.

2.3   Inferred species evolution and lineages

The majority of taxonomic studies have been solely descriptive or for biostratigraphic purposes, with few attempts at the interpretation of tempo of origins and evolutionary relationships of species. This is unfortunate for biostratigraphy as well as for understanding evolutionary patterns, since lineages are among the best possible data for biostratigraphy. Among the exceptions are the trepostome *Prasopora* in Middle Ordovician deposits of a large area centered on the Great Lakes and the cheilostome *Metrarabdotos* in Atlantic Cenozoic deposits. The inferred species lineages in *Prasopora* are rather simple, involving few species that exhibit a recognizable trend for about 15 million years. Like most in the fossil record, the apparent lineage in *Prasopora* may represent ecological displacement rather than direct lineages. The lineages determined for *Metrarabdotos* are set in a context of much greater stratigraphic and geographic control, and therefore seem much more firmly demonstrated.

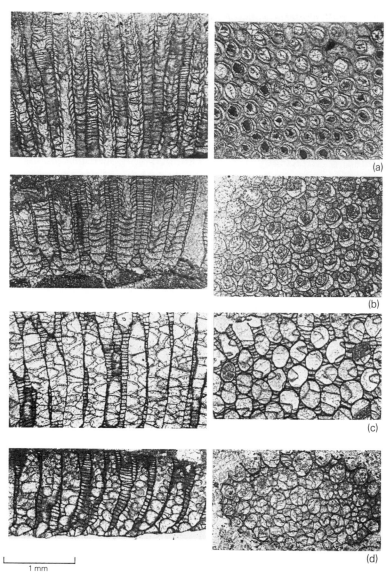

1 mm

Figure 2.5 Longitudinal and tangential sections of species of *Prasopora* to show the evolution of increased space between cystiphragms and reduction in transverse length of zooecial wall that they occupy: (a) *P. discula*, (b) *P. falesi*, (c) *P. selwyni*, (d) *P. shawi*. See Figure 2.6 for stratigraphic distribution and inferred evolutionary relationships.

2.3.1 Prasopora

Prasopora is a Middle Ordovician trepostome characterized by typically free-living, discoidal to hemispherical colonies, thin-walled autozooecia and kenozooecia, curved plates (termed cystiphragms) within the autozooecia, and very small skeletal rods. Although the biological function of cystiphragms is unknown, the distribution of their various morphologies through the Middle Ordovician in North America has been well documented.

In the middle Middle Ordovician, *Prasopora* is represented by *P. discula*, in which the cystiphragms are very closely spaced and in many instances completely encircle a reduced central or subcentral open space within zooecia (Fig. 2.5a, McKinney 1971). Throughout most of the upper Middle Ordovician (Trentonian), only one species is known, *P. falesi* (Sparling 1964, Ross 1967, Marintsch 1981). This species is characterized by overlapped, relatively closely spaced cystiphragms that wrap most of the way around the perimeter of zooecial chambers (Fig. 2.5b). However, in the uppermost Trentonian beds, three additional morphologies abruptly appear (within the limits of stratigraphic resolution), all apparently derived from *P. falesi* by increased spacing between cystiphragms and a decrease in the transverse length of wall that they occupy (Figs. 2.5c, d). The new morphologies appear almost at the same stratigraphic level (Fig. 2.6) in localities scattered across several hundred kilometers. Long periods of morphological stasis in *Prasopora* species interrupted by comparatively very short intervals of morphological change (or species replacement) are more consistent with the model of punctuated equilibrium (Eldredge & Gould 1972, Gould & Eldredge 1977) than with gradual evolutionary change.

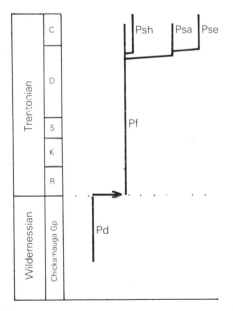

Figure 2.6 Stratigraphic distribution and inferred evolutionary relationships of species of *Prasopora* in the Middle Ordovician of eastern North America: Pd, *Prasopora discula*; Pf, *P. falesi*; Psh, *P. shawi*; Psa, *P. sardesoni*; Pse, *P. selwyni*. (Based on data in Ross 1967 and McKinney 1971.)

2.3.2   Metrarabdotos

The predominantly erect bilaminate, *Metrarabdotos*, is represented abundantly in Eocene to Neogene deposits on both sides of the Atlantic (Cheetham 1968) and survives to the present. There is only one reported fossil or Recent occurrence outside the Atlantic Ocean (i.e., Pacific Ocean off Panama). Therefore, all species encountered through the Cenozoic can be claimed with fair confidence to have evolved within the Atlantic and not to be invaders descended from some unknown non-Atlantic stock.

The morphology and transatlantic distribution of *Metrarabdotos* was documented in detail, and an evolutionary history was proposed for the genus two decades ago (Cheetham 1968). More recently, specimens from relatively complete Neogene sections in the Dominican Republic provided the opportunity to examine species evolution and interrelationships even more closely, and to re-evaluate the originally proposed evolutionary history (Cheetham 1986b, Cheetham & Hayek 1988). Both the original and contemporary studies are summarized here to give historical perspective on this as an example of the progressively more refined interpretations that can derive from sustained and careful work on good material distributed through reasonably complete stratigraphic sections.

Metrarabdotos is morphologically complex, and at least 23 characters of autozooids and heterozooids were used for species taxonomy in 1968. Five of the characters were based on 11 measured variates as combinations ("size") or as ratios, including "shape". In the restudy, 46 measured, counted, and coded morphologic characters were used.

Convergence and parallelism, even in the most diagnostic characters, caused difficulty in the original study in interpretation of evolutionary lineages, even though the fossil record was relatively good. Standard procedures of numerical taxonomy produced five phenetic clusters of species based on the 23 characters, each having 2–5 possible states. The five phenetic clusters in the 23-dimensional morphological field of *Metrarabdotos* yielded very small overlap (Fig. 2.7). The five clusters were interpreted as subgeneric clades, whose species make coherent distributional patterns in time and space, with *M. (Rhabdometra)* as the most primitive group (Fig. 2.8).

Evolutionary trends as exemplified among and within the subgenera can be seen in avicularia, gonozooids, and autozooidal orifices (Fig. 2.9). Size increases in normal avicularia, with loss of special avicularia, gave rise to the predominantly New World *M. (Uniavicularium)* in the Miocene. In the other three derived subgenera, the special avicularia became enlarged over their original size in *M. (Rhabdometra)*. General reduction in costules (small ribs) of gonozooids occurred in both New World and Old World lineages, although more rapidly in the latter, and adventitious avicularia were lost from gonozooids in most New World species. Denticulation of autozooidal orifices evolved from a single median denticle in Eocene and

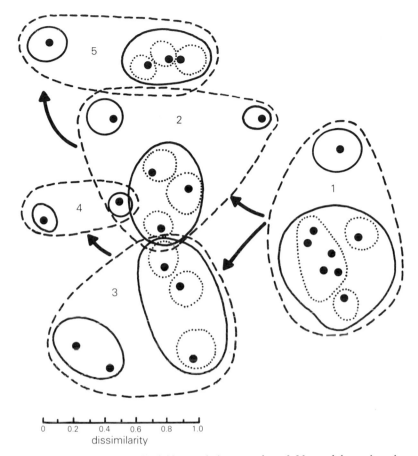

Figure 2.7 Morphologic field of 30 population samples of *Metrarabdotos*, based on 23 characters and projected onto a two-dimensional field. The field is divided into five subgenera: 1, *M. (Rhabdometra)*; 2, *M. (Biavicularium)*; 3, *M. (Porometra)*; 4, *M. (Metrarabdotos)*; 5, *M. (Univavicularium)*; each point represents one or more population sample. Dashed lines encircle subgenera, solid lines encircle species, and dotted lines encircle subspecies. Phylogeny is represented by heavy arrows. (From Cheetham 1968, courtesy of the Smithsonian Institution Press.)

Oligocene species to two lateral denticles in all Pleistocene and Recent species, with intermediate tridenticulate Miocene and Pliocene species when the two end-member states overlapped.

These and other evolutionary trends were most tangled in the New World subgenus *M. (Biavicularium)*, which lacked a single, unique character state exhibited by all populations. Cheetham (1968) defined *M. (Biavicularium)* on ten character states, of which each species displayed at least six and none exceeded nine, whereas no species in another subgenus possessed more than three. As defined, all subgenera were polythetic to some degree, and *M. (Biavicularium)* and *M. (Univavicularium)* defined the extremes.

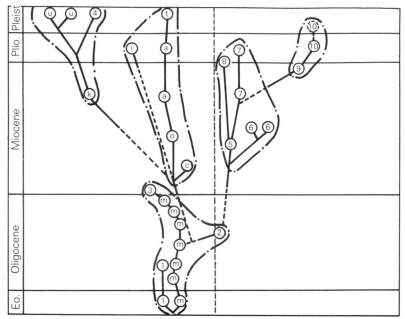

Figure 2.8 Inferred phylogenetic relationships of the five subgenera of Metrarabdotos, based on numerical affinities and stratigraphic distribution. Species identified by letters appear again in Figure 2.10: a = *M. (B.) auriculatum*; c = *M. (B.) chipolanum*; k = *M. (U.) kugleri*) 1 = *M. (B.) lacrymosum*; m = *M. (R.) micropora*; o = *M. (B.) colligatum*; t = *M. (B.) tenue*; u = *M. (U.) unguiculatum*. Numbers identify other species. (After Cheetham 1968, courtesy of the Smithsonian Institution Press.)

Closely sampled populations from good, continuous exposures in the Dominican Republic, estimated to be over 60% complete for 4.5 million years of Miocene and Pliocene deposition, have allowed a reassessment of evolution in *Metrarabdotos* (Cheetham 1986b). The populations include 12 distinct ($P < 0.001$ in the final discriminant analysis) species whose evolution is characterized by long intervals of stasis, punctuated by rapidly splitting lineages. Within each species, the rate of overall morphologic change does not differ significantly from zero and varies in direction randomly about the species' mean throughout its range in the section (Fig. 2.10). These conclusions are sustained also after character-by-character reanalysis of the data (Cheetham 1987).

The original interpretation of evolution within *Metrarabdotos* had been based on more widely spaced populations, and gradual evolution was inferred, often involving subspecies. "The chronocline I thought was represented by the *M. (B.) tenue* 'subspecies' is perhaps the most conspicuous casualty of the restudy, which shows that the supposed cline members largely overlap each other in time. Eldredge & Gould were certainly right about the danger of stringing a series of chronologically isolated populations together with a gradualist's expectations" (Cheetham, personal communication 1986).

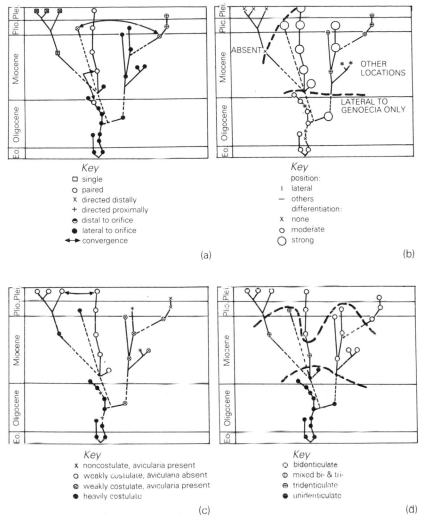

Figure 2.9 Inferred evolution of (a) ordinary avicularia, (b) special avicularia, (c) gonoecia, and (d) oral denticulation in *Metrarabdotos*, based on the phylogeny represented in Figure 2.8. (From Cheetham 1968, courtesy of the Smithsonian Institution Press.)

The restudy of the Dominican Republic *Metrarabdotos* resulted in a re-evaluation of its phylogeny, the rethinking of which led to an important test of the relative merits of cladistic- and stratophenetic-based interpretations of phylogeny (Cheetham & Hayek 1988; see also Sokal (1985) for a more general review). The first step in the re-evaluation involved generation of models of phylogenies. The effects of various sources of morphologic changes and of stasis on the ability to correlate morphology with time of its origin were then assessed. Morphologic distances were measured along the branches of the model, i.e., they are **patristic distances**

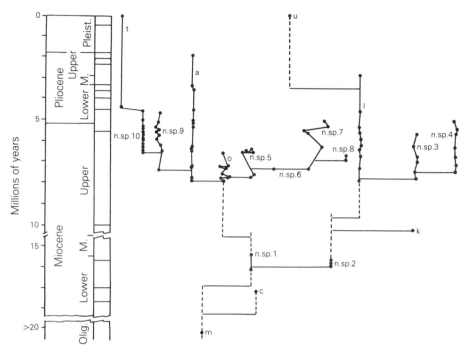

Figure 2.10 Patterns of overall morphologic stasis and punctuational origin in species of *Metrarabdotos*, based on discriminant analysis of characters shown by series of population samples in the Dominican Republic. Numbers and letters identify species as in Figure 2.8. (From Cheetham 1986b, courtesy of the Paleontological Society.)

rather than **phenetic distances**, the latter being direct comparisons of morphology rather than tracing it back along the branches. The sources of morphologic differences examined include evolutionary rates (a) within species and (b) at speciation events; (c) the duration of stasis within species; and (d) random fluctuations in morphology through time within species.

The results indicate that even the most extreme variations between evolutionary rates within species and at speciation events have only minor effects on the morphology–time correlation. In other words, whether evolution is gradual or punctuated, it has little effect on correlation of morphology and time within clades. However, if morphology within species fluctuates more than the differences in average morphologic distance between species, the correlation becomes unpredictable. Also, increased stasis and overlap in temporal ranges can cause a decline to nonsignificance in the morphology–time correlation. Nonetheless, multiple simulations of phylogenies that encompass actual values for evolutionary rates within and between species, morphologic variability within species, and durations of stasis and ancestor–descendant overlaps in *Metrarabdotos*, have overall correlations that are better than the isolated effect of stasis and overlap. A correlation of 70% ($r = 0.7$) was empirically determined as the threshold for agreement between time of origin and patristic distance within a phylogenetic tree.

The 46 measured, counted, and coded characters of the 18 species of *Metrarabdotos* were reduced to 17 canonical discriminant functions for construction of a stratophenetic tree. The tree was rooted in *M. micropora* (the oldest species included) and was constructed by connecting the oldest populations of each species to the stratigraphically proximate species that was closest phenetically. Parametric and nonparametric correlations between patristic distances and times of origin are 0.71 and 0.77, respectively. Therefore, the stratophenetic tree is above the threshold for the "introspective" ability to recognize a statistically realistic phylogeny.

Comparison of species distributions in Figures 2.8 and 2.10 show that the pertinent clades originally defined by Cheetham survived the detailed stratophenetic re-evaluation pretty well. The greatest discrepancy is *M. (Biavicularium)*, whose species are in part redistributed. The reason for the extremely polythetic characterization of *M. (Biavicularium)* becomes apparent: it was originally a polyphyletic ("garbage can") grouping that more detailed study has clarified.

Cladistic trees were constructed, using relative time of occurrence to set the polarity of ancestor–descendant states for each character. A large number of trees were constructed, using the full complement of 46 characters, progressively eliminating the least diagnostic down to a residue of the three most diagnostic characters, and various types of variable-reducing multivariate procedures. The very best temporal correlations that could be generated were 0.45–0.55, which were for the tree based on the nine most diagnostic characters and the tree based on rotated discriminant scores. (The latter was more internally consistent in accurately grouping conspecific populations.) Although the ability of the tree based on rotated discriminant scores is highly significant ($P < 0.001$) in its ability to correlate morphology with time, the low r is substantially below the threshold that a punctuated pattern can produce, and it contains extensive temporal misplacement of populations within species (Cheetham & Hayck 1988).

In the case of *Metrarabdotos*, the stratophenetically produced phylogenetic tree contains fewer errors than do even the best cladistically based phylogenies. It appears that a well-exposed, relatively complete, and thoroughly studied fossil record is necessary for realistic phylogenetic reconstruction of a clade without biochemical evidence, and that prevalent cladistic procedures can at best produce less accurate approximations.

The study of *Metrarabdotos* clearly demonstrates the value of detailed study of fossil sequences in studies of evolution. Thus far it has demonstrated independent parallelism and convergence of characters, tempo of evolution, apparent "decoupling" of instantaneous evolution within species from species origins, the value of samples from closely spaced stratigraphic intervals, and provided a real world test of the relative value of different taxonomic methods. None of these results would have come about or been as rigorously demonstrated without extensive use of various multivariate statistical methods. However, this – which is by far the most exhaustive – study of bryozoan evolution has led inevitably to other questions, while not attempting to answer others that require living animals. The biological functions of avicularia, differentiation of several types of avicularia within a colony, costulation of gonozooids, and orificial denticulation are all unknown. Therefore, although the evolutionary patterns in *Metrarabdotos* are among the best documented for any living or fossil clade, their possible adaptive significance is unknown.

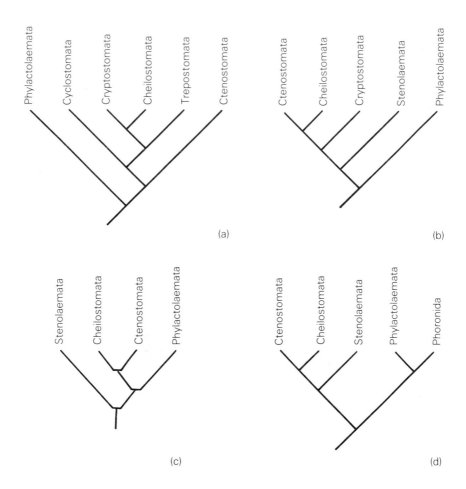

Figure 2.11 Cladograms: (a) after Bassler (1953), (b) after Jebram (1973), and (d) after Larwood & Taylor (1979) and Mundy *et al.* (1981); and a dichotomous classification that implies evolutionary relationships: (c) after Cuffey (1973). Each is based on a different combination and weighting of characters and yields radically different interpretations of relationships.

2.4   Untested phylogenies

Convergence in zooidal morphology and colony form, and the conservatism of polypide organization throughout the stenolaemates and gymnolaemates have resulted in ignorance of phylogenies at higher taxonomic levels within the phylum. Very different results have been obtained when different characters are included or emphasized (Fig. 2.11). For example, presence of an ascus and structure of the frontal wall in cheilostomes had been considered by many specialists earlier in this century as subordinal characters, but with the inclusion of more characters in classification schemes, they are now considered to constitute polyphyletic grades of evolution (Cheetham & Cook 1983). There is well-founded concern that the post-Paleozoic cyclostomes are polyphyletic, incorporating several relicts of the predominantly Paleozoic stenolaemate orders (Boardman 1984).

Detailed knowledge of phylogenetic relationships throughout the Bryozoa is needed before many questions on their evolutionary history can be sensibly addressed. The snare of analogies, particularly among the morphologically simple taxa, is so potentially misleading that phylogenies based solely on morphology will not suffice. The most promising avenue is to determine molecular similarity (Sibley & Ahlquist 1983) among the living bryozoans and then to re-evaluate their morphologies for renewed comparison with extinct taxa.

3 Growth and form

Colonies grow by proliferation of new zooids and extrazooidal tissues. Colony form is determined by the position of new zooids with respect to those formed earlier, and their shape, orientation, and rate of addition in each portion of the colony. Patterns of zooidal addition determine the range of colony morphologies that potentially and actually exist, and therefore constitute both a phylogenetic and architectural constraint on colony form. In this chapter we describe the ways bryozoans grow into different colony morphologies.

3.1 Encrusting growth

New zooids are added to encrusting colonies along part or all of the colony margin. In the first case, growth results in runner-like colonies if the zones of active growth are narrow and divide at regular intervals. If these zones are broader, irregular in width, and divide at irregular intervals and widths, then irregularly lobate sheets result. In contrast, growth around the entire margin produces more radially symmetrical forms. Frontal budding may also increase the thickness (height) of lobate or radially symmetrical encrusting sheets.

3.1.1 Runners

Runners grow as uniserial or narrow multiserial branches that increase by bifurcation at branch tips or by lateral branching at places along the branch. Runners extend rapidly (§ 5.3) and, depending on the rate of branching, may explore growth in several directions along the substratum.

   The geometric constraints and consequences of highly patterned runner growth may be seen in two species of Jurassic *Stomatopora* that are formed of dichotomous branches (Fig. 3.1, Gardiner & Taylor 1982). Within fossil colonies, the bifurcation angle averages 155° for the first but diminishes to and stabilizes around 75° by the fourth and subsequent bifurcations (Fig. 3.2). Gardiner & Taylor ran a series of simulations of colony growth in which the angle of bifurcation remained constant across all generations, or was systematically increased or decreased with successive generations (Fig. 3.3). They found that any constant angle of bifurcation from 1° to 180° results in branch convergence after growth of three sets of links of equal length, and that progressively increasing the angle of bifurcation only accelerates this convergence. However, progressively decreasing the angle of bifurcation

Figure 3.1 Portion of a colony of *Stomatopora dichotomoides*, Middle Jurassic (Bathonian), Baunton, Gloucestershire, England. (Photography courtesy of P. D. Taylor.)

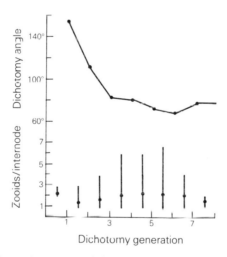

Figure 3.2 Plots of branching characteristics of *Stomatopora bajocensis* (d'Orbigny) and *S. dichotomoides* (d'Orbigny). The upper curve plots mean dichotomy angle against dichotomy generation as numbered from ancestrula, and the lower graph plots mean and range in numbers of zooids per internode. (From Gardiner & Taylor 1982, courtesy of E. Schweizerbart'sche Verlagsbuchhandlung.)

delays branch convergence in proportion to the rate of decrease. Branch convergence in fossils apparently caused termination of one of the branches and loss of any descendant branches that might have derived from it. Thus, delay in convergence due to astogenetic decrease in bifurcation angle should have increased colony size, zooid number, and reproductive potential.

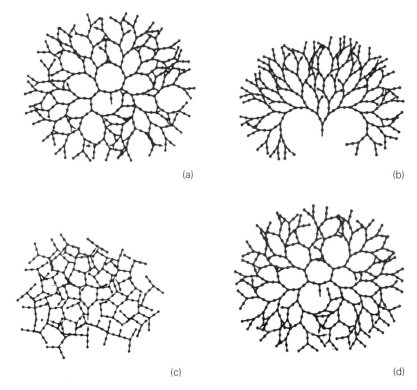

Figure 3.3 Simulations of (a) a specific colony of *Stomatopora*, and of artificial uniserial colonies in which (b) the bifurcation angle remains constant, (c) the bifurcation angle constantly increases, and (d) the bifurcation angle decreases exponentially; note the similarity of (a) and (d). (After Gardiner & Taylor 1982, courtesy of E. Schweizerbart'sche Verlagsbuchhandlung.)

3.1.2   Single and multilayered sheets

Encrusting sheets add new zooids along continuous peripheral margins. Commonly, new zooids are added at similar rates around the entire periphery of the colony, resulting in radially spreading buttons or sheets, although variable rates of growth or the presence of obstructions result in colonies with irregular outlines. Stronger polarity in growth rates of encrusting sheets results in belt-shaped colonies such as develop in *Membranipora membranacea* on fronds of the alga *Laminaria hyperborea* (Ryland & Stebbing 1971). Most of these colonies acquire a growth orientation that is roughly proximal down the algal stipe before they reach 3.5 cm diameter, and 80% of all colonies – some up to more than a meter in length – grow as broad belts straight down the frond.

Encrusting sheets growing on limited substrata may overgrow themselves where actively growing edges meet, or where a vigorous edge encounters an older, senescent portion of the colony. In this manner multilayered colonies of either stenolaemates or cheilostomes may form through self-overgrowth. Self-overgrowth also occurs in stenolaemates in which zooids, either by differential growth or, more typically, as a result of damage to their neighbors, extend above the adjacent apertures. The overtopping zooids establish a new multizooidal budding zone that extends laterally over the surrounding surface. This occurred widely in Paleozoic trepostomes where local areas either became senescent or were damaged (Bigey 1981).

Multilayered growth by various forms of frontal budding is common among Cenozoic and Recent cheilostomes. The best studied example is *Schizoporella unicornis* (Banta 1971, 1972), in which the body cavities above frontal calcified shelves of parental zooids balloon upwardly as the intrazooidal origins of buds. Vertical walls of new zooids are upward extensions of vertical walls of parental zooids, and interserial communication organs between adjacent "vertical" lineal series are established where exterior cuticle between series is locally dissolved. Lidgard (1985a, b) also found that multilayered growth can result from coalescent multiserial budding and from nonlineal multiserial budding (Fig. 1.21). In *Celleporaria magnifica*, among others, chaotic frontally budded regions include a mixture of the latter two types of budding.

Multilayered growth is typical of mound-like and nodular sheet-like to massive colonies. It has developed commonly in species that grow on limited substrata, such as pagurid-inhabited gastropod shells (stenolaemates – Fisher & Buge 1970, Palmer & Hancock 1973, Taylor 1976, cheilostomes – Osburn 1950, p.54, Cook 1964) and other small substrata that develop as rolling stones (cheilostomes – Rider & Enrico 1979). It is also the predominant mode of growth among abundant species on rock walls and reefs (§ 7.5).

3.2 Erect growth

Erect colonies have much more complex form than do encrusting colonies, and their growth is correspondingly more difficult to describe. To aid description, and to provide possible insight into the geometric and adaptive bases of different erect growth forms, there has been increasing use of models and simulations of growth of these organisms. Results have shown that very complex three-dimensional forms can be generated by specifying only a few parameters.

3.2.1 Unilaminate and bilaminate sheets

Among the erect growth forms, unilaminate and bilaminate sheets seem to exhibit the least regularity in their development. They grow essentially as encrusting sheets that have escaped from the substratum.

The growing edge of many encrusting species may lift off the substratum to form erect unilaminate sheets (Cheetham 1971, Jackson & Buss 1975). This may occur when the growing edge contacts an obstruction or the limit of the substratum, or for no apparent reason. Bilaminate sheets may be produced either haphazardly or as part of a regular, closely coordinated process. For example, where the cheilostome *"Hippodiplosia" insculpta* encrusts algae and gorgonian stems, the advancing edges of a single colony may meet where they have wrapped entirely around the substratum (Nielsen 1981). This results in the growing edges lifting up off the substratum and continuing growth back-to-back at a more or less coordinated pace, producing an erect bilaminate sheet. Nielsen noted that growing edges of different colonies of *"H." insculpta* can also lift off of the substratum where they meet and encrust one another to produce bilaminate sheets, and he noted at least one interspecific example formed by a colony of *"H." insculpta* and another of *Thalamoporella californica*.

Growth of bilaminate sheets in stenolaemates begins where a ridge-like wall forms within the colony. This wall becomes a median wall within a confluent multizooidal budding zone (Boardman 1983). Two oppositely facing budding zones are centered on the median wall and generate new zooids back-to-back as the median wall advances. The advance of the growing edges of both sides of the bilaminate sheet are therefore exactly coordinated by the common median wall.

3.2.2   Arborescent bilaminate (adeoniform) colonies

Colonies composed of multiserial bilaminate branches, with zooids opening on both flattened sides, and that are continuously and rigidly calcified from base of attachment to branch tips, are termed **adeoniform**, following the usage originally intended by Brown (1952). Growth of colonies of most adeoniform species follows regular and precise species-specific rules of growth that fluctuate little as the colonies grow (Fig. 3.4, Cheetham *et al.* 1980, 1981, Cheetham & Hayek 1983, Cheetham 1986a). The principal results are significantly reduced overlap and variation in spacing between the growing tips of branches, which further increases both the potential growth and surface area of colonies.

Growth of adeoniform colonies involves four processes:

(a) continuous branch elongation by budding of new zooids;
(b) branch widening, which results from multiplication of zooid rows;
(c) branch multiplication by division of the growing tip; and
(d) branch thickening proximal to the growing tip, which does not extend the colony into space but contributes to colony strength (§ 8.3.2).

Bifurcation angle and mean length of branch links (segments between successive bifurcations) are constant throughout the cheilostome colonies that have been studied, although link length varies substantially and

1 cm

Figure 3.4 Four colonies of the Recent adeoniform cheilostome *Cystisella saccata* (Busk) arranged as a growth series. (Photographs courtesy of A. H. Cheetham.)

unevenly in some Paleozoic adeoniform species. In most species, two branches extending from a single bifurcation grow at unequal rates, whether they be from the first bifurcation above the substratum or one formed in later growth stages. Moreover, fully formed branch links extending from a single bifurcation are unequal in length, the inequality forming a characteristic ratio for each species examined. Each branch path consists of the sum of the links from the proximal end of a colony to a specific branch tip,

$$\overline{L}_t = (1/G_t)(NaA + NbB)$$

and the total path length within a colony is

$$T_t = naA + nbB$$

where $\bar{L}_t$ is mean path length; G_t is number of branch tips; Na and Nb are the summed number of links of length A and B, respectively, in each path; T_t is total path length; and na and nb are the total number of links of length A and B within a colony. The relative rates of increase in $\bar{L}_t$ and T_t reflect colony bushiness and can be important in adaptive morphology, for the former measure is proportional to height and the latter to surface area. The wide variation in $T_t:\bar{L}_t$ with growth of Cenozoic adeoniform cheilostome species (Fig. 3.5a) reflects differences in height and bushiness among them. The most highly significant differences are between bifurcation angle and link length. Interspecific differences in link length seem to be major contributors to the differences in $T_t:\bar{L}_t$, for when the two values are divided by mean link length ($\bar{l}$) the differences in ratio largely disappear (Fig. 3.5b).

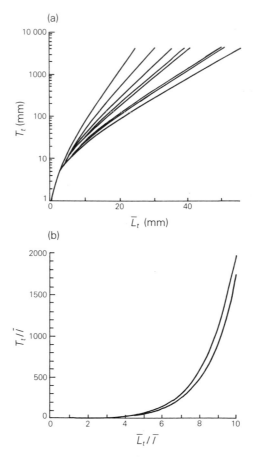

Figure 3.5 (a) Increase in total path distance (T_t) against mean path length ($\bar{L}_t$) for eight Recent adeoniform cheilostomes, calculated from branching properties, and (b) the same after removal of differences in mean link length ($\bar{l}$). (From Cheetham *et al.* 1980, courtesy of Academic Press.)

Figure 3.6 Crossover of converging adeoniform branches (point a) generated by twist at bifurcations proximal to the point of convergence. (After Cheetham & Hayek 1983, courtesy of the Paleontological Society.)

Variation in angle of bifurcation as measured in the plane of the parent branch affects the overall shape of a colony. Relatively low angles of bifurcation cause an increase in colony height, while higher angles cause an increase in colony radius. Overall colony radius is maximum in models where the angle of bifurcation is 90°. In some Paleocene, Oligocene, and Recent adeoniform species, the angle of bifurcation is greater than 60°, even though this results in branch loss at later growth stages when the lowest branches approach the substratum. There has apparently been selection for such high angles of bifurcation during the Cenozoic, for there has been a statistically significant ($P < 0.002$) trend through time towards the 90° bifurcation angle (Cheetham 1986a).

Bifurcations in adeoniform cheilostomes – and in almost all the bilaminate arborescent stenolaemates – occur in the plane along which the two sheets of zooids join. If the plane remained perfectly flat, the median branches would converge into the same space after the third bifurcation and before the fourth could develop, just as in uniformly bifurcated models of encrusting runners (§ 3.1.1). However, the newly developed branches at each bifurcation are twisted with respect to the ancestral branch. As a result, the first pair of converging branches cross over past one another and continue independent growth (Fig. 3.6). The crossover distance in living species averages about five times the initial branch thickness and increases with both the angle of twist and the angle of bifurcation; but it is only the twisting at bifurcations that makes the difference between three-dimensional bushes and perforated sheets.

Bifurcation angle, link lengths, and angle of twist do not change with growth, so that branches eventually begin to converge and to interfere with one another. This sets a limit to growth stage and size of colony, which is approximately $3(A + B)$ and is defined as Growth Stage 3 (Fig. 3.7). Growth Stages 1 and 2 are $(A + B)$ and $2(A + B)$, respectively.

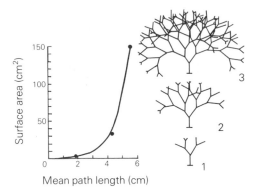

Figure 3.7 Diagrammatic representations of adeoniform growth stages 1–3 and their position on the curve of mean path length (which approximates colony height) against total surface area of one side of branches. (After Cheetham & Thomsen 1981, courtesy of the Paleontological Society.)

Increase in the twist angle causes an increase in the colony's radius perpendicular to the plane of its stem with respect to the radius in the plane of its stem. Regardless of the degree of twist, as colonies grow they continue to increase this ratio, which reaches a maximum at or near Growth Stage 3. The overall colony radius, and also the average nearest-neighbor distance of branch tips at Growth Stage 3, are at maximum if the angle of branch bifurcation is 90° and the angle of twist is 50°. Thus the Cenozoic increase in angle of branch bifurcation has also contributed towards increasing the overall colony radius and the average distance between branch tips at Growth Stage 3 in modern species.

As a colony grows, the average nearest-neighbor distance among branch tips begins a steady decline at the first branch crossover, but there is an even greater decline in minimum distance, which results in increased variability of spacing with growth. The space between branch tips is not evenly distributed at Growth Stage 3, the coefficient of variation reaching 50–60% in both real and simulated colonies, so that it is not all available to accommodate branch growth. This high variation in spacing of branch tips at later growth stages contributes to their interference. The slightly lower level of variation in recent species over that in Paleocene and Oligocene species ($\chi^2 = 4.0$, $P < 0.05$) means that there has been a phylogenetic decrease in branch interference.

Decrease in variability of nearest-neighbor spacing of branch tips is strongly polyphyletic. It appears to be due to whatever advantage accompanies reduced irregularity in spacing between branch tips, which is in turn the result of adhering more closely to a regular growth pattern.

Arborescent growth allows exponential increase in surface area (Fig. 3.8). In adeoniform colonies, branch width and thickness of the growing tips are such that, by Growth Stage 3, surface area is about two times that of an

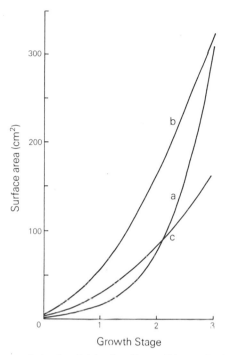

Figure 3.8 Surface areas of simulated (a) adeoniform, (b) massive hemispherical and (c) circular, encrusting sheet-like colonies. Areas are mean adeoniform link length squared. Radii of (b) and (c) equal the mean branch length of (a). For branching characteristics of (a) see Cheetham & Hayek (1983). (From Cheetham & Hayek 1983, courtesy of the Paleontological Society.)

encrusting sheet with the same diameter, although before Growth Stage 2 an encrusting sheet of equal diameter has the greater surface area. Massive hemispherical colonies and adeoniform colonies at Growth Stage 3 that have equal diameter also have approximately equal surface areas, but the arborescent colonies require only about 2–3% the skeletal volume of massive hemispherical colonies, and thus have, presumably, considerable savings in investment in growth.

3.2.3 Arborescent unilaminate colonies

Unilaminate arborescent bryozoans (Fig. 3.9) are composed of closely spaced, narrow branches that are usually arrayed in well-developed flat or curved sheets. Each branch and intervening space is generally 1 mm or less wide, collectively forming a slotted or perforated sheet. Colonies tend to have similar shapes and apparently follow grossly similar rules of growth as do other colonial suspension feeders (gorgonians, hydrozoans, and grapto-lites) with narrow branches closely arrayed in planar sheets (McKinney 1981b). Such planar arrangements of branches are unique among colonies in

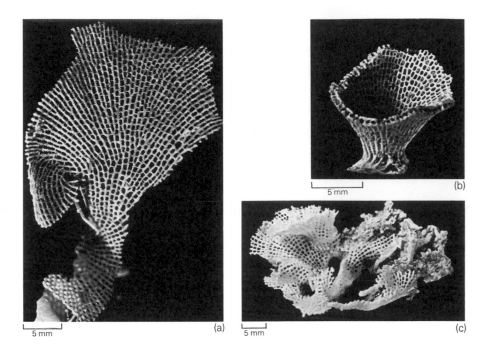

Figure 3.9 Variations in form of unilaminate arborescent colonies of fenestrate steno-laemates, including (a) expanding fan-shaped fenestellid, (b) conical *Unitrypa*, and (c) complexly folded fenestellid. (From McKinney 1981b, courtesy of the Paleontological Society.)

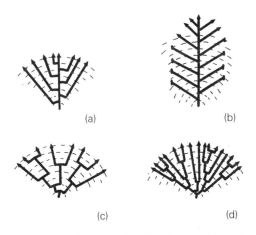

Figure 3.10 Models of fan-shaped growth of uniformly spaced branches within a plane: (a) continuous growth of main branch and pinnae, (b) continuous growth of main branch and discontinuous growth of pinnae, (c) & (d) continuous growth of bifurcated branches. Concentric spacing between branch origins is equal in (a), (b), and (c), but is half the distance in (d).

their very narrow variance in width and spacing of branches. Although transport of materials along branches and problems of structural support may also contribute to their organization, the restriction of this type branching system to colonial suspension feeders suggests that filtration is a critical selective factor in its development (Chs. 6 & 8).

The simplest unilaminate colony form is a wedge-shaped, regularly expanding, planar fan that may be either pinnate or dendroid (Fig. 3.10). Form and texture of pinnate colonies are determined by

(a) distance between the origins of lateral branches on each side of the main branch;
(b) apposition or alternation of lateral branches;
(c) angle between lateral branches and main branch; and
(d) whether growth is continuous, forming an expanding fan as in Figure 3.10a, or is discontinuous, perhaps forming a roughly parallel-sided fan where growth of proximal branches has stopped as in Figure 3.10b.

Form and texture of dendroid colonies, whose branches increase by bifurcation, are determined by

(a) the arc or angle subtended by the growing margin as measured from the colony origin,
(b) centering of each branch between its two neighbors, and
(c) the critical lateral distances between three neighboring branches at which the central branch will bifurcate.

Relationships between length of the colony from origin to growing edge, length of the growing edge, and surface area in uniformly expanding planar fans and cones are quite regular (Fig. 3.11). The number of branches has a direct linear relationship to length of the growing edge. Therefore branch bifurcations, which are the means of increase in branch number, accumulate distally in direct proportion to length of the growing margin (Fig. 3.11) and to the spacing of branches within the fan (Fig. 3.10c, d).

Branch origins in pinnate colonies are highly predictable; they are concentrated along the main branch, and lateral branches have small probabilities of giving rise to descendants. In contrast, dendroid unilaminate colonies tend to have bifurcations scattered across the growing margin. The increased number of branches at their growing margins results in a distal decrease in the probability of bifurcation of any given branch per increment of increase in fan length (Fig. 3.11, line d).

Most unilaminate arborescent colonies do not grow as simply expanding planar fans or cones. More common are fans and cones that expand distally at variable or exponential rates and complexly folded sheets (Fig. 3.9a, c). In some of these more complexly shaped colonies, branch bifurcations are quite obviously concentrated in certain regions, such as along the left margin of the fenestellid colony illustrated in Figure 3.9a. In this colony, the margin adjacent to the prolific bifurcations has spread and recurved over some of its

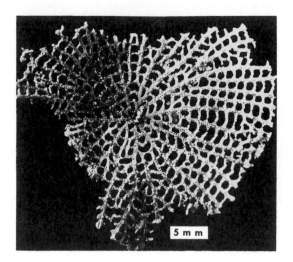

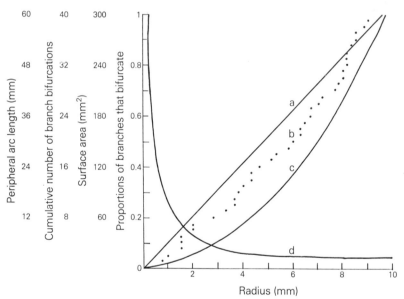

Figure 3.11 Radius of a specimen of the fenestrate *Septopora* (top) plotted against (a) peripheral arc length, (b) cumulative number of branch bifurcations (plotted as points), (c) surface area, and (d) proportion of branches bifurcating within an area of increase in radius equal to the distance between the first and second bifurcations from the colony origin. (From McKinney 1981b, courtesy of the Paleontological Society.)

more proximal parts. This example gives insight into the geometry of branching in regularly spiralled colonies that evolved from fan-shaped ancestors, which has occurred in the Paleozoic fenestrates (*Archimedes*) and in modern cheilostomes (*Bugula*) (Fig. 3.12).

2 mm   (a)   (b)
                                      5 mm

Figure 3.12   Erect spiralled colonies of (a) *Bugula turrita* and (b) *Archimedes Intermedius*, both characterized by whorls of narrow unilaminate branches.

Spiralled *Bugula turrita* colonies can be subdivided into a series of dendroid planar fans (Fig. 3.13) that originate from bifurcations along the axial helical margin of the colony. Serial surfaces cut through the heavily calcified axial margins of *Archimedes* skeletons demonstrate that its colonies are organized in the same general way.

The spiralled form can be modelled solely by establishment and main-tenance of a bifurcating, helical margin in addition to the growth of the radiating simple planar fans (McKinney & Raup 1982). The configuration of the bifurcating, helical margin is determined by

(a) the radius of the helical path,
(b) the rate of climb of the helix (CLIMB),
(c) the radial angle between bifurcations along the helical path, and
(d) the angle between the axis of the helix and the branches that diverge from it (ANGLE).

The diverging branches each serve as the beginning of a planar branch system that is bounded by similar proximal and distal branch systems.

The several parameters of the models of spiralled colonies may be varied to produce a hypothetical morphospace within which the observed morphol-ogies of species of *Archimedes* and *Bugula* fall. For example, Figure 3.14 illustrates models that result from varying CLIMB against ANGLE. The observed morphologies in *Archimedes* and *Bugula* show a narrow range in

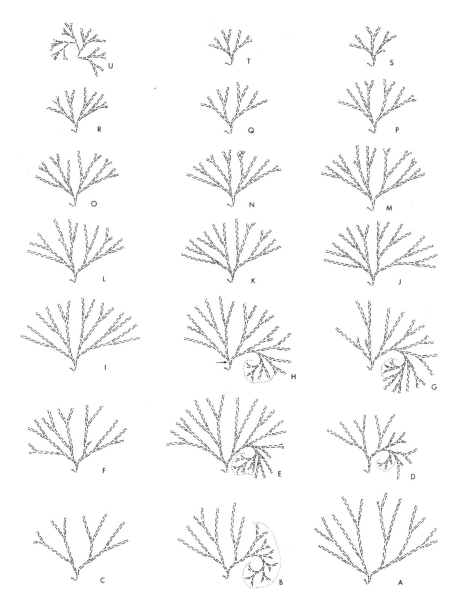

Figure 3.13 Diagrammatic representation of branch systems of a colony of *Bugula turrita*, with each dash representing a single zooid. The proximal branch system is at lower right, and the distal tip of the colony is at upper left; each row from right to left constitutes a whorl of the tuft. The helical axis of the colony is constituted by the proximal pair of zooids in each branch system. (From McKinney 1980, courtesy of Society of Economic Paleontologists and Mineralogists.)

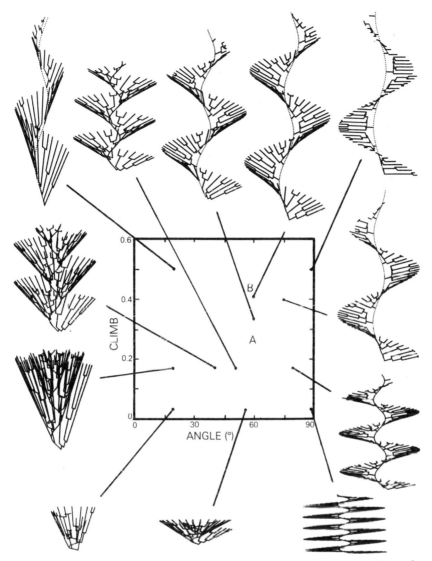

Figure 3.14 Hypothetical morphospace defined by varying rate of translation of axis (CLIMB) and angle between colony axis and lateral branch systems (ANGLE) with constant branch spacing. Approximate positions occupied by *Archimedes intermedia* (A) and of *Bugula turrita* (B) are plotted. (After McKinney & Raup 1982, courtesy of the Paleontological Society.)

ANGLE but a substantial interspecific variation in CLIMB. The interesting questions of *why* the actual morphologies are distributed as they are have yet to be answered by more than preliminary hypotheses.

Arborescent unilaminate colonies vary much more in overall form than do adeoniform colonies. This greater variability appears to result from more freedom in location of branch bifurcations and lengths than occurs in the adeoniform colonies. Some species of unilaminates (cones and spirals) have species-specific colony forms, while others are more indeterminate, with the ultimate form and size dependent in large part on unpredictable clustering of bifurcations and branch growth (Fig. 3.9c).

Among the colonial suspension feeders that form planar sheets, the active feeders (bryozoans and graptolites) are unilaminate and have a very narrow range of branch sizes and spacings, whereas branch organization, sizes, and spacing are more variable in the passive feeders (gorgonians and hydrozoans). Although colony geometry in all arborescent unilaminate bryozoans can be described in terms of branch spacing and position of bifurcations that add new branches, the stimuli for these attributes are at present unknown. Nevertheless, widespread similarities in branch size and spacing, and the range of colony shapes and sizes among distantly related arborescent unilaminate bryozoans suggest a common function, which is discussed further in Chapter 8.

3.3   Free-living growth

Free-living stenolaemates were common from the Ordovician to Permian and grew by lateral growth over fine-grained sediments or by developing a concentric, potentially rolling habit. Concentric self-overgrowth and potentially rolling habit has also developed in post-Paleozoic cheilostomes. However, the most common free-living bryozoans are cap-shaped (**lunulitiform**) cheilostomes. Larval metamorphosis typically occurs on grains that are smaller than the eventual adult colony size.

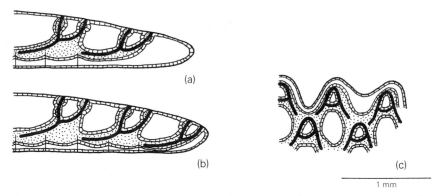

Figure 3.15   Idealized sections through a free-living cupulariid cheilostome. (a) Longitudinal section cutting colonial space from which an autozooid will be partitioned at colony margin on right, (b) longitudinal section cutting skeletally defined vibraculum at colony margin on right, (c) tangential section through autozooids, vibracula, and marginal colonial space. Brick pattern represents epithelium, heavy black lines represent primary skeleton, stippled pattern represents secondary skeleton. (After Håkansson 1973, courtesy of Academic Press.)

Nonlineal budding has been characteristic of the free-living cheilostomes (Tavener-Smith & Williams 1972, Håkansson 1973, Lidgard 1985b). New zooids are partitioned from the ring of colonial space by rapid local proliferation of secreting epithelium situated low on the distal sides of the existing zooids. The newly forming skeleton eventually curves up to contact and fuse with the frontal cuticular wall (Fig. 3.15). Basal, lateral, and transverse walls all lack cuticle, are interior, and their growth divides new zooidal spaces from the undivided colonial space around the periphery, which is a nonlineal budding zone. When the colony grows beyond its initial substratum, the basal wall becomes separated from the basal cuticle by a space lined with epithelium on both surfaces. The space may be undivided, or various combinations of transverse and longitudinal cuticles form at some distance behind the growing edge and connect the basal skeletal wall and the basal cuticle. This anchors the basal cuticle to the skeleton. As the free-living colony grows peripherally, the more proximal skeleton continues to thicken on its underside, often completely engulfing the original substratum.

3.4 Rooted growth

Specialized adaptations for living rooted in sediments have evolved in several groups of cheilostomes and ctenostomes. These include a variety of forms that resemble various erect and free-living colonies. These rooted taxa, however, have turgid, heterozooidal or extrazooidal, cuticle-encased tubes that extend from the proximal or basal regions of the colonies and attach to grains of sediment (Fig. 9.14, Cook & Lagaaij 1976, Cook 1981, Cook & Chimonides 1981a, b). Other rooted forms are specially adapted for life on sediments, including numerous top-shaped, globular, conical, or stellate colonies of only a few millimeters width and length.

Calcified parts of top-shaped zoaria are not attached to any substratum, making interpretation of earliest astogeny inferential even for modern colonies, since early colony growth has not been directly observed. Rather than producing a single ancestrula, larval metamorphosis apparently produces a group of primary zooids, most of which are autozooids but with one, or perhaps more, differently shaped, rootlet-bearing kenozooid(s) (Fig. 3.16). Orifices of the autozooids are oriented towards the substratum, from which growth of the rootlet lifts the colony. Budding adds new zooids on the side of the colony opposite the kenozooidal apex. Thus the "distal" growth direction of the colony is opposite the orificial ends of autozooids, which contrasts with all other bryozoans.

In top-shaped colonies of the families Conescharellinidae and Orbituli-poridae, the frontal wall has expanded at the expense of all the other wall regions. Marginal pores around the flattened hexagonal, orifice-bearing portion of the wall (Fig. 3.16) are inferred to be the loci from which frontal buds originate. Autozooids formed in successive series are progressively

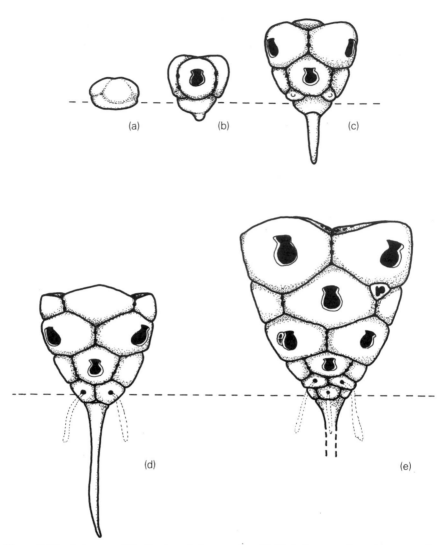

Figure 3.16 Astogeny of idealized, rooted, conescharellinid cheilostome. Larval metamorphosis (a) produces several primary zooids (b), including one or more rootlet-bearing kenozooids (bottom). Continued growth of the colony (c)–(e) adds progressively larger autozooids at the top of the colony and additional rootlet-bearing kenozooids at the proximal end. Note that orifices of autozooids are orientated proximally with respect to the direction in which they are added to the colony. (After Cook & Lagaaij 1976, courtesy of the British Museum of Natural History.)

larger in most top-shaped colonies of these families, so that a region of repetition of zooidal morphology never develops. Additional gradients in zooidal morphology may develop simultaneously with, or subsequently to, the initial zone by frontal budding from exposed frontal walls of zooids that are proximal to the distal edge of the colony.

Top-shaped colonies of genera in the family Mamilloporidae develop by more typical distal and disto-lateral budding. In these the distal growth direction of the colonies and orientation of autozooids correspond. Differences in orientation and growth of these colonies compared with the Conescharellinidae and Orbituliporidae are probably related to opposite polarity of ancestrular complexes that are developed at completion of metamorphosis.

4 Growth forms as adaptive strategies

Bryozoans have always grown in many shapes, but these can be reduced to a few basic forms (Fig. 4.1). Bryozoan colonies differ in the ways zooids are arranged relative to each other (**uniserial** versus **multiserial**) and in their principal directions of growth, which can be horizontal (encrusting) or vertical (erect). In uniserial colonies, one zooid typically buds another end-on, thereby forming single or loosely branching chains (for convenience we include biserial forms here). In multiserial colonies, zooids bud others laterally as well as distally to form more-or-less continuous surfaces of zooids instead of chains.

There are five basic growth forms that include the great majority of Recent and fossil Bryozoa characteristic of hard substrata:

(a) uniserial (and biserial) encrusting,
(b) multiserial encrusting,
(c) massive multiserial encrusting,
(d) multiserial erect, and
(e) uniserial (and biserial) erect.

Two additional multiserial growth forms, highly specialized for life on sediments, are free-living and minute rooted species. These are discussed separately in Chapter 9.

In this chapter we compare the distributions of the five most common growth forms among four major extant groups of marine Bryozoa. Next we develop a simple model to describe the apparent costs and benefits of these five forms relative to risks of mortality from different causes. These deductions lead to testable predictions about the zooidal characteristics and environmental distributions of species with different growth forms. These predictions are generally sustained for both living and fossil bryozoans, which suggests that risks of mortality have selected for different growth forms in much the same ways throughout the Phanerozoic. However, the relative intensity of different factors has apparently changed greatly, as evidenced by massive temporal shifts in environmental distributions of the different growth forms.

4.1 Frequencies of growth forms among major taxa

A compilation was made of all bryozoan species reported in 47 ecological surveys and taxonomic monographs for the Atlantic basin from the Arctic Ocean to South Africa, and for the western Mediterranean. Effort was made

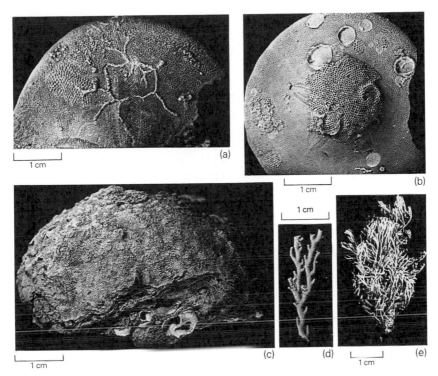

Figure 4.1 Basic growth forms of bryozoans: (a) uniserial encrusting *Stomatopora*; (b) multiserial encrusting *Conopeum*; (c) massive multiserial encrusting *Stylopoma spongites*; (d) multiserial erect *Hornera*; (e) uniserial erect *Crisia*.

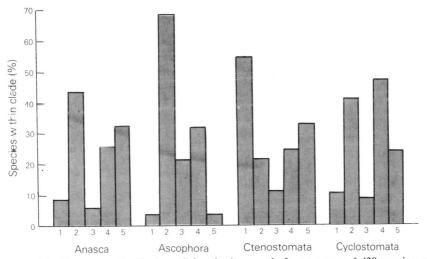

Figure 4.2 Frequency distribution of five basic growth forms among 1,430 species of bryozoans in the Atlantic Ocean. Growth forms: 1 = uniserial encrusting, 2 = multiserial encrusting, 3 = massive multiserial encrusting, 4 = multiserial erect, 5 = uniserial erect. Based on 47 references available from Jackson on request.

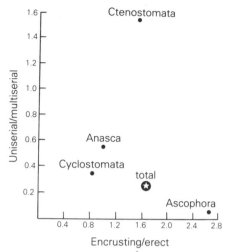

Figure 4.3 Comparison of the proportions of uniserial to multiserial and of encrusting to erect species within the four major extant taxa of marine bryozoans. The star indicates the average proportion for all groups combined.

to avoid duplication of species using available synonymies. This produced 3,300 records for a total of 1,430 species including 1,146 cheilostomes (419 anascans, 44 cribrimorphs, 683 ascophorans), 99 ctenostomes, and 185 cyclostomes. Growth form was determined from the papers or from original taxonomic descriptions; depth and substrata were recorded when available (Jackson, unpublished).

Distributions of the five forms for all taxa combined are shown in Figure 4.2. More than half the species are multiserial encrusting, and the next most abundant are multiserial erect. The least common form is uniserial encrusting. Multiserial species are about four times commoner than uniserial, and encrusting 1.7 times commoner than erect.

The distribution of the five forms is strikingly different between major taxa (Figs. 4.2, 4.3). Ctenostomes are the most uniserial of the groups, followed by anascans, whereas cribrimorphs and ascophorans are overwhelmingly multiserial. Cribrimorphs (not shown in Figs. 4.2, 4.3) are exclusively encrusting and ascophorans very largely so, whereas anascans and cyclostomes include as many or more erect as encrusting species. Excluding the uncommon cribrimorphs, anascans and cyclostomes are the most similar to one another in colony form, ctenostomes and ascophorans the most dissimilar.

4.2 Growth form model

What is the ecological and evolutionary significance of this variation? One approach to this question is to deduce the adaptive significance of different growth forms as strategies to reduce risk of mortality due to known

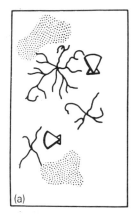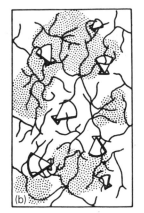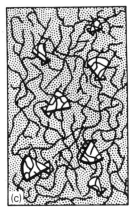

Figure 4.4 Example of refuge location of the uniserial cyclostome *Stomatopora* sp. on settlement panels. Schematic drawings show distributions over time of *Stomatopora* (lines), bivalves (shells), and encrusting organisms that are superior competitors to *Stomatopora* but not to bivalves (stippled): (a) after seven months, (b) after 14 months, (c) after 26 months. *Stomatopora* colonized all kinds of surfaces but zooids survived only on shells where they were protected from overgrowth. (From Buss 1979, courtesy of the Systematics Association.)

ecological processes (Jackson 1985). Such a model artificially simplifies the problem to manageable proportions and leads to a series of testable hypotheses regarding the importance of bryozoan colony shapes to variations in their zooidal morphology, life history, and distribution.

The model used here considers risks of mortality for different growth forms living on hard substrata in terms of differences in their modes and directions of growth (Jackson 1979a, Coates & Jackson 1985).

Uniserial colonies are highly directional forms whose growth increases their chance of locating particular types of favorable habitats (spatial refuges) on an heterogeneous terrain (Fig. 4.4, Buss 1979). However, zooids in a chain lack neighbors on all sides for common protection against physical damage or enemies. Thus, the probability that any zooid will die in a given interval is high, but the probability that some small group of zooids somewhere in the colony will find a refuge and survive is also high. In contrast, multiserial growth suggests an increased commitment to survival in the immediate vicinity of larval attachment. Thus, multiserial colonies employ what might be termed a confrontational strategy in which survival is more dependent on maintenance of the genetic individual as a single integrated colony, or as a few integrated colonies, than in uniserial clones that may be highly fragmented. In this case the probability that any zooid will die is typically much less than in uniserial forms, but the chances of locating refuges by growth are lower as well.

Whether uniserial or multiserial, encrusting growth sets no special mechanical constraints for support of the colony. Thus growth is potentially infinite, although in practice limited by the boundaries of substrata, interactions with other organisms, and the vagaries of the physical environment. The costs of such structural freedom are greater vulnerability to all the potentially harmful processes that occur more intensively on the substratum – including sedimentation, overgrowth, and some forms of predation. In contrast, erect growth places stringent mechanical demands on skeletal systems for support and attachment (Ch. 8, Cheetham 1971). The potential benefits include greater access to resources above the substratum, escape from harmful processes restricted to the bottom, and enormous increase in surface area and biomass, and thus feeding and reproduction, for a given area of habitat.

4.3   Zooidal characteristics and colony form

Bryozoan zooids are the basic units of construction from which colonies are made (Ch. 1). Zooids vary considerably in size, shape, spacing, design, function, and integration, all of which should be related in some way with colony form.

4.3.1   Predicted zooidal characters

Survival of a genetic individual of a uniserial species often depends on survival of one or a few isolated zooids. In this case, chance of further survival should increase with zooidal size, implying selection for size increase. Zooids should be elongate or widely spaced to maximize location of spatial refuges. Integration of zooids and incidence of zooidal polymorphism should be low to increase chances that any surviving zooid can feed or reproduce. Moreover, when polymorphism occurs, different types of zooids should be distributed uniformly throughout the colony to increase the chances that all functions might survive.

Survival of a multiserial colony rarely depends on a single zooid. Potential size of a colony is largely independent of zooidal size so there should be no selection for zooidal size increase, at least so far as risk of mortality is concerned. In contrast, integration of zooids and incidence of zooidal polymorphism should be high, with clearly developed separation of functions (different kinds of polymorphs) within different regions of colonies, to increase chances of survival for the colony as a whole.

For both uniserial and multiserial growth, biomechanical influences on zooidal characters should differ greatly between encrusting and erect colonies. Because there are no special structural problems associated with encrusting growth, zooidal characteristics, and their expression within a colony, should be free to vary widely in response to biological factors affecting them. In contrast, mechanical constraints on erect growth, particularly that of arborescent species, sets strict limits on the distribution of types of zooids and special supporting structures independently of their potential advantages or disadvantages in biological interactions.

Intermediate growth forms between uniserial and multiserial, or encrusting and erect, are common, as in the fan- or ribbon-like colonies of *Electra* or *Steginoporella* (Figs. 5.9a, 5.10) that are intermediate between uniserial "runners" and multiserial "sheets". Zooidal characteristics of such colonies should be intermediate.

4.3.2   Observed zooidal characteristics

Zooidal morphology, spatial arrangement, and polymorphism are generally as predicted by the risk model in both cheilostomes and cyclostomes (Coates & Jackson 1985).

Results for cheilostomes are summarized in Table 4.1 and Figure 4.5. Elongation and spacing vary as predicted by the model, with maximal values for uniserial species. However, zooidal size is least for these species because their zooids are extremely narrow. As the model predicts, the highest incidence of polymorphism is for multiserial species, most of which have both ovicells and avicularia, and lowest for uniserial species, most of which have only one.

There are also data on the relation of colony form to integration of zooidal feeding behavior (Winston 1978, 1979). As predicted, integration is greatest for multiserial colonies, and least for uniserial ones (Fig. 4.6), but there is no significant difference between encrusting and erect colonies.

Table 4.1 Variations in zooidal size, elongation, spacing, and incidence of polymorphic zooids in relation to colony form among recent and early Tertiary (Paleocene, Eocene) cheilostomes. Recent data from Harmer 1926, 1957; Ryland and Hayward 1977; and Hayward and Ryland 1979; fossil data from Canu and Bassler 1933; Cheetham 1963, 1966, 1971. Size is the geometric mean of zooidal length times width $\sqrt{L \cdot W}$; elongation is L/W; and spacing is the mean distance between all orifices surrounding a zooid. Polymorphism was scored 0 = no ovicells or avicularia, 1 = avicularia or ovicells present, 2 = both kinds of polymorphs present. F-ratios and probabilities for 1-way analysis of variance of parameter values by colony form.

Colony form	Size		Elongation		Spacing		Polymorphism	
	recent	fossil	recent	fossil	recent	fossil	recent	fossil
uniserial								
unilaminate	0.38	0.52	2.69	2.80	0.57	0.59	1.50	0.50
multiserial								
unilaminate	0.48	0.45	1.49	1.53	0.44	0.41	1.47	1.31
multiserial								
multilaminate	0.45	0.56	1.33	1.48	0.41	0.43	1.93	1.00
multiserial erect	0.43	0.44	1.86	1.90	0.38	0.43	1.79	1.16
uniserial erect	0.43	0.40	3.01	2.12	0.52	0.43	1.40	1.29
mean	0.45	0.45	2.00	1.76	0.45	0.43	1.54	1.22
number of species	462	138	462	138	414	127	522	143
F-Ratio	3.8	1.5	90	8.3	7.4	1.5	8.9	1.5
probability	.004	.21	.000	.000	.000	.21	.000	.21

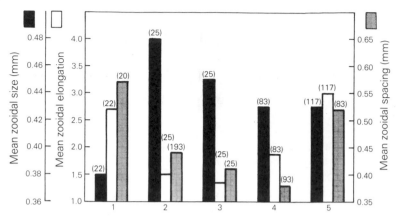

Figure 4.5 Size, elongation, and spacing of zooids of recent species of cheilostomes from Britain (Ryland & Hayward 1977, Hayward & Ryland 1979) and the Indonesian region (Harmer 1926, 1957). Zooidal size is the geometric mean of zooidal length times width, elongation is the length divided by the width, and spacing is the average distance between the orifices of all zooids surrounding a zooid. Zooids may grow adjacent to one another or widely apart, so that spacing of neighboring zooids may be similar to or greater than zooid size. Growth forms are as in Figure 4.2. Values in parentheses are numbers of species for each form.

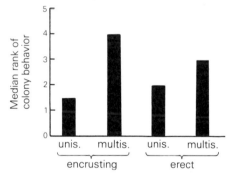

Figure 4.6 Median rank of colony feeding behavior of species described by Winston (1978) in relation to their colony form. Higher values indicate higher integration of feeding among zooids.

Four cheilostome faunas from the Paleocene and Eocene (Canu & Bassler 1933, Cheetham 1963, 1966, 1971) were examined to determine to what extent zooidal characteristics predicted by the risk model were established at the beginning of the Tertiary radiation of the group (Coates & Jackson 1985). Results are very different from the Recent (Table 4.1): only zooidal elongation varies significantly between growth forms of Paleocene and Eocene cheilostomes, and the other three parameters do not show even a suggestion of the modern pattern. This increase in zooidal differentiation between growth forms suggests considerable, sustained selection for more than 50 million years. Similarly, the large overall increase in incidence of polymorphism (compare last two columns of Table 4.1) suggests a compar-

edge   surface

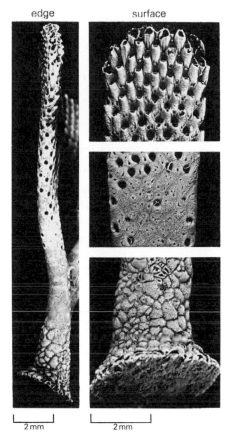

2 mm   2 mm

Figure 4.7 Extrazooidal thickening of the basal region of the rigidly erect arborescent cheilostome *Cystisella saccata* which increases resistance to breakage by the colony while decreasing the numbers of feeding zooids. (From Cheetham & Thomsen 1981, courtesy of the Palcontological Socicty.)

ably important increase in integration of zooids, as had been previously indicated from somewhat different evidence for the entire span of cheilostome evolution (Fig. 1.9, Boardman & Cheetham 1973).

Perhaps the best example of interdependence of growth form and zooidal characters is that of rigidly erect cheilostomes which first appeared in several independent lineages during the late Cretaceous (Cheetham 1971, 1986a). Binding, thickening, and joining together of zooids all depend on the existence of specific zooidal morphotypes, without which growth of erect, rigid cheilostomes could not occur. An obvious example is the benefit of frontal calcification of zooids associated with the extrazooidal thickening of branches towards the bases of arborescent colonies (Figs. 4.7 & 8.8, Boardman 1954, Cheetham *et al.* 1981, Cheetham & Thomsen 1981). More subtle is the relation between thickness of zooids at the growing tips of rigidly erect colonies and their inferred modulus of rupture, i.e., resistance

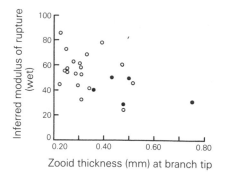

Figure 4.8 Relation between thickness of zooids and their inferred modulus of rupture in Recent (open circles) and fossil (closed circles) rigidly erect bilaminate arborescent bryozoans. (Data courtesy of A. H. Cheetham.)

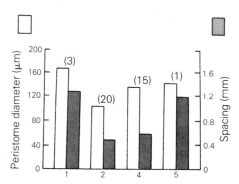

Figure 4.9 Size and spacing of zooids for 33 Recent species of tubuliporine cyclostomes from the western Mediterranean (Harmelin 1976). Because of zooidal overlap, size was estimated from peristomal diameter, and zooidal elongation could not be measured. Spacing was determined as for cheilostomes. Growth forms as in Figure 4.2; values in parentheses are numbers of species for each form.

to bending (Ch. 8), as shown for 24 Recent and fossil species in Figure 4.8. Modulus of rupture decreases significantly with increasing zooidal thickness ($r_s = -0.52$, $P < 0.01$).

Results for 33 species of Mediterranean tubuliporine cyclostomes are shown in Figure 4.9. Uniserial species have larger and more widely spaced zooids, as predicted by the growth form model.

4.4 Environmental distributions

Regardless of growth form, any sessile animal can survive where there is adequate substratum and food, the weather is favorable, competitors are absent, and potentially harmful processes like storms, sedimentation, or predation (often lumped together by ecologists as "disturbance") do not occur. With competition, or extreme or unpredictable variation in disturbance, however, different growth forms may be varyingly excluded or diminished in importance.

4.4.1 Predicted distributions

According to the risk model, uniserial colonies should be favored over multiserial ones at high levels of disturbance because they are more likely to have encountered a spatial refuge from disturbance. Moreover, growth requirements are less substantial than for multiserial forms, so that uniserial colonies are more likely to be able to grow large enough to reproduce before a calamity strikes. Also at high levels of disturbance, encrusting colonies should be favored over erect ones because they can reproduce sooner (less growth required), and because encrusting colonies are harder to break or eat.

In contrast, at low levels of disturbance, multiserial colonies should be favored over uniserial colonies because they are better competitors for space. They can also completely cover a substratum and thereby prevent invasion by foreign larvae. Similarly, erect colonies should be favored over encrusting ones because they can get closer to food suspended in the water and, in so doing, reduce access to food by encrusting colonies beneath them, just as erect table corals inhibit those beneath them by shading (Stimson 1985).

The predictions of the model are easy to test in principle, but not in practice, because the distribution of intensity of competition and different forms of disturbance on the sea floor are not well known and are not independent. Important forms of physical disturbance include waves and currents, sedimentation, and extreme fluctuations in salinity or temperature, all of which decrease in intensity with increasing depth (Sverdrup et al. 1942). Their impact, however, may vary depending on a host of other factors. For example, deleterious effects of strong water movements should vary with the size or stability of the substratum; if it is small, slight movements may turn it over, whereas if the substratum is large or firmly attached this will not happen, and the water movements may even be beneficial as a source of extra food. Similarly, flexible or short-lived substrata, like fronds of algae and seagrass blades, are more easily moved about or destroyed than attached scleractinian corals. Indeed, the very abundance of suitable substrata, and thus that of bryozoans themselves, may depend upon the local physical hydrographic regime (Eggleston 1972a, Holme & Wilson 1985). Lastly, extreme variations in factors like salinity and temperature are much more likely in confined coastal areas than along open coasts.

Important forms of biological disturbance are predation and bioturbation. Like physical disturbance, both of these processes decrease with depth. The abundance of fishes, sea urchins, limpets, and other important grazers of hard substrata drops off considerably within 100–200 m of the surface (Lang 1974, Jackson & Winston 1982). Similarly, deposit-feeding animals that turn over sediments, and thus may disturb substrata suitable for bryozoans, like shell fragments or shells, are less abundant and turn over less sediment below a few hundred meters (Appendix 1 in Thayer 1983). Regardless of depth, the effects of both these processes may vary locally with the relative accessibility or susceptibility of habitats to these organisms.

In summary, it seems very likely that the total amounts of disturbance that may affect bryozoans fall off greatly with increasing depth, with increasing size or stability of the substratum, and away from estuaries and enclosed bays. Thus, according to the risk model, the ratios of uniserial to multiserial and encrusting to erect species should decrease in the same directions.

4.4.2 Observed distributions with depth

In our Atlantic sample, numbers of bryozoan species and all attached

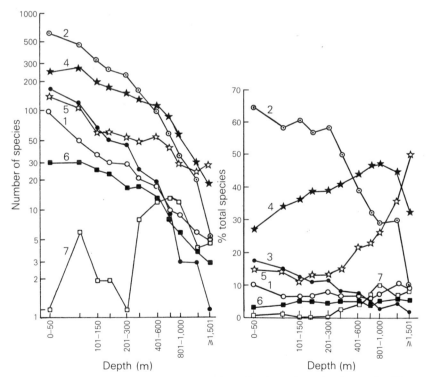

Figure 4.10 Distribution of growth forms by depth for the entire Atlantic sample of Recent bryozoans for which depth data were available. (a) Numbers of species of each growth form within each depth interval, (b) proportion of species of any growth form as percentage of the total number of species in each depth interval. Percentages add to more than 100% because some species adopt more than one growth form. Growth forms 1–5 as in Figure 4.2; 6, free-living; 7, minute rooted.

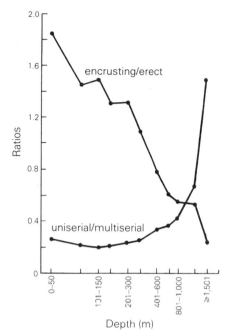

Figure 4.11 Ratios of uniserial to multiserial and of encrusting to erect bryozoan species as a function of depth, from the survey of Recent Atlantic bryozoans.

growth forms fall off precipitously with depth, except for the minute rooted forms discussed in Chapter 8 (Fig. 4.10a). Values of the two ratios are illustrated in Figure 4.11. The ratio of encrusting to erect species falls off as predicted, with erect species becoming more diverse only below about 400 m. However, the uniserial to multiserial ratio shows little change in shallow depths, and then increases sharply below 1,000 m, which is contrary to prediction. Examination of the relative frequencies of the different growth forms with depth (Fig. 4.10b) shows that this contrary result is due to an overwhelming predominance of uniserial erect species at great depths where they comprise half the fauna. In contrast, multiserial erect species show a more complex distribution, increasing in relative diversity with depth to about 1,000 m, and then falling off again to about the same percentage as at the surface.

Some insight into these patterns emerges from information on the attachment and construction of erect species which may be:

(a) cemented or rooted,
(b) inflexible or flexible (independent of jointing), and
(c) unjointed or jointed.

Between the surface and 150 m, there is a sharp, more-or-less parallel increase in the proportions of cemented, inflexible, and unjointed species (Fig. 4.12). This is followed by a gradual decline to 1,000 m, and then an

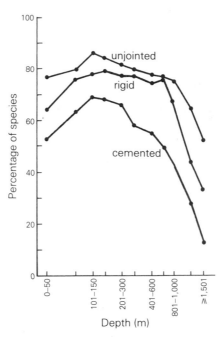

Figure 4.12 Percentages of erect bryozoans that are cemented, rigid, or unjointed as a function of depth from our Atlantic survey. Percentages of rooted, flexible, or jointed species are the mirror image of values shown.

even steeper decline below. Moreover, this pattern largely reflects that of multiserial erect species, while the pattern for uniserial species is much less pronounced. Cemented attachment, inflexible construction, and lack of joints combine to make a rigid structure compared with colonies that are rooted, flexible, or jointed. This rigidly erect construction is less frequent in shallow waters, presumably due to problems of breakage by water movements (Holme & Wilson 1985) and predators (Vance 1979). On the other hand, its subsequent precipitous decrease with depth (beginning for cemented species between 300 and 400 m) must be due to other causes, most probably decreased availability of suitably stable substrata, and perhaps frequent shifting of sediments from turbidity flows. The impression is that the deep sea is not so environmentally stable as is often suggested, at least with regard to substrata for sessile attachment.

There are too few quantitative abundance data for living colonies to compare with these patterns. Trawl and camera surveys in the North Sea (Dyer *et al.* 1983) and English Channel (Holme & Wilson 1985) have revealed abundant populations of foliaceous erect species, such as *Flustra foliacea* and *Pentapora foliacea* (Fig. 4.13), but other smaller or less conspicuous bryozoans were not examined. The latter study is particularly valuable for clearly showing the complex interacting effects of water movements, in this case tidal currents up to 160 or even 250 cm sec^{-1}, and

Figure 4.13 Several colonies of the large foliaceous erect cheilostome *Pentapora foliacea* at about 50–55 m depth in the English Channel. Width of area is 67 cm. Bottom is stable, hard, and exposed to very strong currents but not to heavy sediment accumulation or scour. (From Holme & Wilson 1985, courtesy of the Marine Biological Association U.K. and Cambridge University Press.)

Figure 4.14 Drawing of a large colony of *Flustra foliacea* from the bottom of the English Channel. (From Stebbing 1971, courtesy of the Marine Biological Association U.K.)

the effects of transported sediments. Stable, hard bottoms exposed to such water movements, but not subject to scour or burial by sediments, are covered by abundant encrusting sponges, bryozoans, ascidians, etc., including large numbers of rigidly erect *Pentapora foliacea* and *Omalosecosa ramulosa* (Fig. 4.13). In contrast, similar bottoms subject to scour and potential for burial lack sponges and rigidly erect bryozoans, but support numerous multiserial erect, flexible colonies of cemented *Flustra foliacea* (Fig. 4.14) and rooted *Vesicularia spinosa*. Foliaceous species also dominate several faunas below 30 to 100 m off New Zealand (Probert & Batham 1979, Bradstock & Gordon 1983) and in the Gulf of Mexico (Canu & Bassler 1928, J. E. Winston, personal communication).

Rigidly erect cheilostomes are abundant below 50 m in the Ross Sea (Fig. 4.15, Winston 1983, Hayward & Taylor 1984) and several coastal bays along the Antarctic coast, where they may comprise one-third or more of all the bryozoan species present, and on the Grand Banks off Newfoundland (Powell 1968). However, they do not appear to be at all common in tropical seas shallower than 100 m, particularly in reef environments (Jackson 1984, unpublished). In contrast, small colonies of rigidly erect species are the predominant growth form among cyclostomes in the majority of habitats in depths less than 100 m in the Mediterranean (Harmelin 1975, 1976). Large, rigidly erect cyclostomes do occur today (Fig. 4.16, Moyano 1973, Schopf *et al.* 1980), but much less frequently than cheilostomes.

4.4.3   Distributions on different substrata

Bryozoans grow on an enormous variety of substrata. To simplify analysis, we grouped these into three categories:

(a) live, mostly soft or flexible – including all algae (except crustose corallines), seagrasses, hydroids, ascidians, etc. – and carapaces or shells of live crustaceans, molluscs, or hermit crabs;
(b) mostly dead, hard substrata – including shell fragments, shells, pebbles, stones, concretions – and living or dead crustose corallines; and
(c) dead surfaces of scleractinian corals.

The reason for separating corals is that they are typically much larger and longer lived than the other hard substrata, and are commonly cemented to the bottom. Stability of the three groups of substrata is inferred to increase in the order presented. Rock walls are even more stable than corals, but could not be included for lack of data.

Distributions of species on these substrata are illustrated in Figure 4.17 for two sets of data. The shaded bars are for species in our entire Atlantic sample, the open bars for cheilostomes from south-east Florida and the Caribbean in depths less than 100 m, for which ecological data are more complete (Winston 1982, 1984a, Jackson 1984). Stability increases from left to right. The pattern for uniserial versus multiserial species is as predicted.

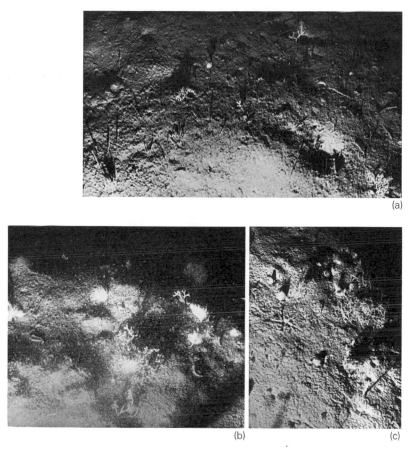

Figure 4.15 Bottom photographs from the Ross Sea, Antarctica, at several hundred meters depth, showing large numbers of rooted, multiserial erect bryozoans. (a) Blade-shaped colonies of *Melicerita oblique* and clumps of cellarinellids, (b) clumps of cellarinellids which form clonally, (c) thicket of cellarinellids in upper right and ophiuroid to left. (From Winston 1983, courtesy of the *Bulletin of Marine Science*.)

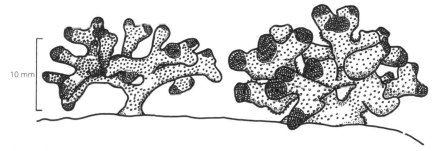

Figure 4.16 Drawing of 2 large colonies of the rigidly erect cyclostome *Heteropora pacifica* from a rock wall at Friday Harbor, Washington. (After Schopf *et al*. 1980, courtesy of the Paleontological Society.)

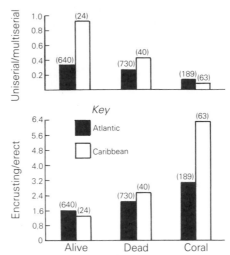

Figure 4.17 Ratios of uniserial to multiserial and encrusting to erect bryozoans as a function of the stability of their substrata which increases from left to right (see text). Solid bars for entire Atlantic survey fauna, open bars for Florida and Jamaica only. Numbers of species in each group are given in parentheses.

Uniserial species comprise almost half the fauna on algae and seagrasses at many sites, and are common on shell fragments and small bits of rubble, but are rare on corals. In contrast, the ratio of encrusting to erect species is opposite that predicted by the risk model. This is due to the great predominance of uniserial erect species on algae and seagrasses, and the rarity of any erect species on corals (note especially the Caribbean data). Predation upon sessile organisms is intense on coral reefs (Sammarco 1980, Jackson & Winston 1982, Jackson & Kaufmann 1986), and multiserial erect species are virtually absent except for a few cryptic habitats. Thus predation appears to be more important than substratum stability for the occurrence of erect species, at least in shallow water (see also Vance 1979).

4.4.4 Distributions with salinity

Table 4.2 contrasts the distributions of taxa in our Atlantic survey with Winston's (1977a) compilation of species reported from waters of diminished salinity world-wide. As predicted by the risk model, the two overwhelmingly multiserial groups, ascophorans and cyclostomes, are the most diminished in brackish waters, while the predominantly uniserial ctenostomes are diminished least. In contrast, there is no relation between the overall ratio of encrusting to erect species in each taxon and their representation in brackish water.

We further examined the relation between growth form and salinity using Winston's (1982) survey of the fauna from south-east Florida. There are 21 species from waters of diminished salinity, and 63 from normal marine

Table 4.2 Taxonomic distributions of species reported from brackish water world-wide (Winston 1977a) and for all species from all environments in our compilation of Atlantic and western Mediterranean bryozoans. Data are numbers of species and percentages in parentheses. Differences in taxonomic proportions are highly significant ($\chi^2 = 108$, $P < 0.001$).

Taxon	Brackish water	All Atlantic environments
Anasca*	92 (45%)	463 (32%)
Ascophora	43 (21%)	683 (48%)
Ctenostomata	52 (26%)	99 (7%)
Cyclostomata	16 (8%)	185 (13%)
total	203 (100%)	1430 (100%)

*Includes cribrimorphs.

waters. As predicted, the ratio of uniserial to multiserial species is extremely high in brackish water – 2.4, whereas it is only 0.28 elsewhere. Contrary to prediction, however, the ratio of encrusting to erect species is lower in brackish water (1.2) than elsewhere (5.6). This higher proportion of erect species in brackish water is of special interest because of the decrease in abundance and diversity of predators in estuaries (Jackson 1985), a pattern which further supports the idea that predators play a major role in determining the relative abundance of erect growth forms in shallow water.

4.5 Growth forms in the fossil record

Growth forms of many fossil bryozoans are different in important details from Recent species (Ch. 8), but they can still be analyzed in the same general way.

4.5.1 Uniserial versus multiserial fossils

Uniserial encrusting bryozoans have been common but not diverse on shells and similar substrata throughout the history of the phylum (Fig. 4.18). Examples include the widespread occurrence of uniserial cyclostomes such as *Corynotrypa* and *Stomatopora* on shell debris from the Ordovician to the end of the Mesozoic (Fig. 4.19, Bassler 1911, Taylor 1979a, b, 1984, 1985a), and the earliest cheilostome, *Pyriporopsis* (Fig. 1.22, Pohowsky 1973, Taylor *et al.* 1981). In contrast, uniserial erect cheilostomes are uncommon as fossils, probably due to problems of preservation of their weakly calcified skeletons. Uniserial erect forms comprise 16% of the 1,430 Recent species in our survey, but only 7% of the Paleocene and Eocene species in Table 4.1. Similarly, uniserial encrusting species are common in Maastrichtian hardgrounds and seagrass deposits, but uniserial erect species are not (Voigt 1981). Consequently, compilation of ratios of uniserial to multiserial fossil species seems fruitless and potentially misleading.

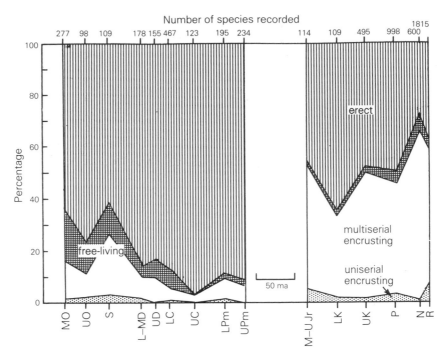

Figure 4.18 Percentage of species in fossil and Recent faunas that are erect, free-living, or encrusting (uniserial or multiserial). Based on 58 references available from McKinney on request.

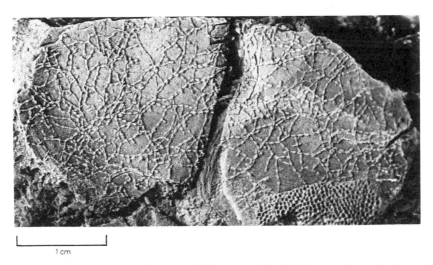

Figure 4.19 Ordovician *Corynotrypa inflata* encrusting the inner surface of a bumastid trilobite molt, Benbolt Formation (Middle Ordovician), Virginia.

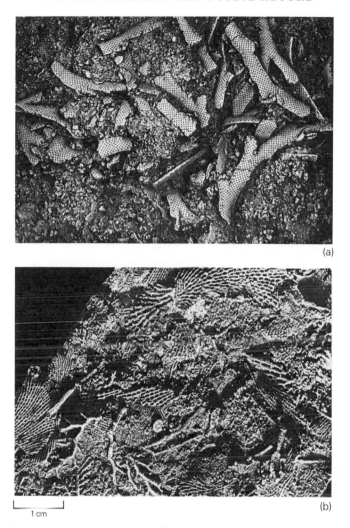

(a)

(b)

1 cm

Figure 4.20 Bryozoan-rich deposits dominated by remains of rigidly erect species. (a) *Coscinopleura digitata* from the Paleocene Vincentown Formation, New Jersey (from Cheetham *et al.* 1981, courtesy of the Paleontological Society); (b) various fenestellids, *Rhabdomeson*, and *Ramiporalia* from the Carboniferous Bangor Limestone, Alabama.

4.5.2 Encrusting versus erect fossils

There was a complete reversal in the diversity of encrusting versus erect bryozoan species between the Paleozoic and post-Paleozoic (Fig. 4.18). Overall, the median ratio of encrusting to erect species for the Paleozoic is 0.10, and for the post-Paleozoic is 0.97 (Mann–Whitney U Test, $P = 0.0009$). Paleozoic faunas are overwhelmingly dominated taxonomically by erect species, and appear to have become increasingly so after the Silurian (Mann's test for trend, $T = 3.20$, $P = 0.007$). In contrast, post-

Paleozoic faunas include, on average, equal numbers of encrusting and erect species and, since the Jurassic, have become dominated by encrusting species, especially in the Neogene.

The predominance of encrusting species in Cenozoic seas is in stark contrast to the impression gained from the best-known fossil bryozoan faunas from the Ordovician onwards. Rigidly erect cheilostomes volumetric-ally dominate the majority of bryozoan-rich deposits of Upper Cretaceous to Recent age (Fig. 4.20a, Canu & Bassler 1933, Cheetham 1963, 1966, 1971, personal communication, Lagaaij & Gautier 1965, Labracherie 1973, Thomsen 1976, Balson 1981, Cheetham & Thomsen 1981, Cuffey *et al.* 1981). Likewise, rigidly erect stenolaemates usually dominate Paleozoic and Mesozoic deposits (Fig. 4.20b, Bassler 1906, Morozova 1961, 1970, Astrova 1965, Walter 1969, Hillmer 1971, McKinney 1972, Kopaevich 1984).

The taxonomic and volumetric dominance of rigidly erect multiserial cheilostomes and cyclostomes on Cretaceous seagrasses is particularly interesting (Fig. 4.21, Voigt 1973, 1981). More than 70 of 130 species found on the Masstrichtian seagrass *Thalassocharis* formed rigidly erect colonies. In contrast, only 6 of 110 species living today on *Posidonia* in the Mediterranean have this morphology (Harmelin 1976); the rest are en-crusting or grow as typically uniserial, flexibly erect colonies.

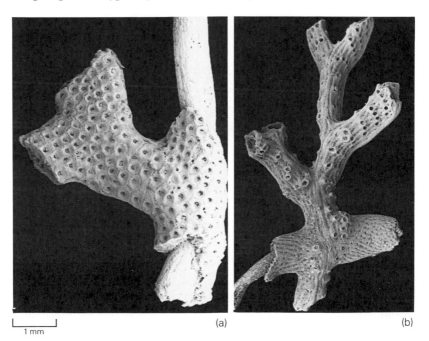

1 mm   (a)   (b)

Figure 4.21 Rigidly erect bryozoans that lived on the Cretaceous seagrass *Thalassocharis*, Maastrichtian, Netherlands. (a) *Onychocella nysti*, (b) *Reteporides lichenoides*, with base encrusting fossil seagrass root. (From Voigt 1981, courtesy of Olsen & Olsen.)

For a variety of reasons, these differences between volumes of rigidly erect colonies in fossil and Recent faunas could be more apparent than real. Ecologists studying living bryozoans have worked primarily in the intertidal zone or at depths less than 20 m (Chs. 5, 7, 9). Deeper water faunas have been investigated primarily by systematists provided with little ecological information, making Winston's (1983) study of inferred patterns of growth, reproduction, and mortality of Antarctic rigidly erect species unusual. Thus, Recent faunas dominated by these forms may be more common than now realized.

On the other hand, many rigidly erect bryozoans probably grow fast and suffer extensive breakage and partial mortality as do the majority of branching corals (Highsmith 1982). If so, great accumulations of fossil skeletal material may suggest abundances of erect bryozoans far greater than they were in life, much like the problem of discarded moults of trilobites. Indeed, much of the sea bottom visible in photographs for Antarctica (Fig. 4.15) seems rather barren of erect bryozoans, even though the photographs were specifically chosen to illustrate these organisms; yet the sediment on which the bryozoans are growing is rich in bryozoan debris (Winston 1983).

There is other evidence, however, that suggests more fundamental differences between ancient and Recent distributions of rigidly erect species. Modern depth distributions of growth forms in shallow water, particularly the upper limit of rigidly erect species, are very different from Paleozoic distributions. Then, these species were the dominant forms on hard substrata, both in diversity and estimated abundance, throughout epicontinental seas (Fig. 4.22, Cuffey 1967, Bretsky 1970, Ross 1970, McKinney 1979, McKinney & Gault 1980, Karklins 1984, McKinney et al.

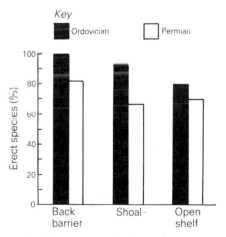

Figure 4.22 Percentage of erect species of bryozoans from different epicontinental environments in the Ordovician and Permian. Depth increases from left to right. (From Ross 1970, Newton 1971, Warner & Cuffey 1973, Simonsen & Cuffey 1980.)

1986b). Many were quite large, commonly reaching 10–20 cm in height or diameter, and occasionally 1–3 m (e.g., Ross 1972, Smith 1981). Similarly, large erect forms dominated shallow level bottoms. Bryozoans such as *Archimedes*, *Fistulipora*, *Tabulipora*, and various Ordovician trepostomes built up enormous populations on lagoonal floors and shoals (Cuffey 1967, Ross 1970, McKinney 1979, 1983, Ettensohn *et al.* 1986). These environments today are dominated by branching corals, or may lack erect animals entirely.

Physical processes in shallow waters are unlikely to have changed radically during the Mesozoic, certainly not enough to push multiserial erect species 100 m deeper, which strongly suggests a biological explanation. The most likely candidate is the dramatic escalation in effectiveness of the grazing, browsing, and gouging machinery of potential predators upon bryozoans throughout the Mesozoic (Vermeij 1977, Steneck 1983). Particularly pertinent are the appearances of sea urchins with well-developed Aristotle's lanterns in the Jurassic and Cretaceous, and of fishes capable of excavating calcareous material in the Eocene. A good example is the sea urchin, *Centrostephanus coronatus*, that feeds preferentially on erect bryozoans and other erect sessile organisms on shallow rock walls off southern California. Consequently, erect animals are rare except when the urchin has been removed experimentally or is naturally absent or rare (Vance 1979). Abundances of all these predators drop off sharply with increasing depth.

The present is not always the key to the past. The presence of rigidly erect colonies, particularly those with radial branches, commonly had been accepted as suggestive of relatively "deep" offshore shelf deposits in studies of Paleozoic bryozoans (e.g., McKinney 1972) until numerous instances providing irrefutable evidence of very shallow, often almost intertidal, occurrences of these forms were recognized. Apparently, shallow shelf environments are more disturbed by predators than in the Paleozoic. Thus, the deeper modern distribution of rigidly erect species conforms in a general way with the predictions of the risk model.

4.6   Growth forms as adaptive strategies

So far we have considered different growth forms as alternative ways of making a living, without explicitly considering species that grow in more than one shape. However, 340, or 24%, of the species encountered in our survey have this potential. Such variability most commonly involves transitions between uniserial encrusting species and either multiserial encrusting species ("runner" to "sheet") or uniserial erect forms ("runner" to "vine"). Also common are transitions between multiserial encrusting and foliaceous rigidly erect growth, which merely involves raising of the growing edge away from the substratum.

Variations in encrusting growth patterns within a species seem readily explicable in terms of plastic responses to changing environmental conditions. The best example is that of *Conopeum tenuissimum*, whose growth changes from predominantly uniserial to multiserial with increasing quantity and quality of food (Fig. 6.1). Another example, perhaps also related to food availability, is the ribbon-like growth of *Steginoporella* sp. under deep corals and its broad fan-like growth under shallow corals (compare Figs. 5.9a, 5.10). Similar variations occur in colony morphologies of rigidly erect species (Harmelin 1973, 1975).

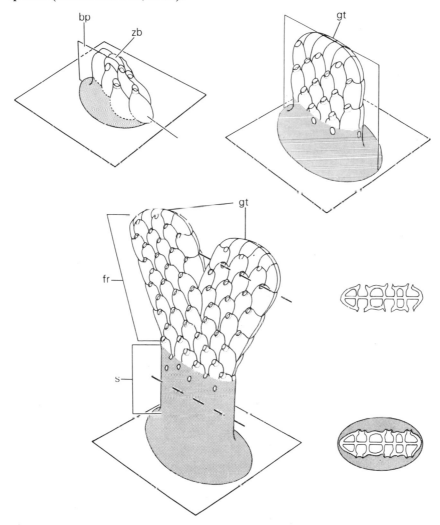

Figure 4.23 Early growth of an idealized rigidly erect cheilostome, showing the invariant transition from encrusting to bilaminate erect growth. bp, bilaminate plane; fr, feeding and reproduction; gt, growing tip; s, support; zb, zooidal bud. (From Cheetham 1986a, courtesy of the Royal Society, London.)

Cheilostomes capable of frontal budding become multilaminar once this process begins, but the timing is so variable that some colonies remain unilaminate throughout their lives, whereas others begin multilaminate growth almost immediately. Early frontal budding may be a response to restrictions to continued horizontal growth, such as the limits of a substratum, or to interference by other species, as in *Antropora tincta* (Buss 1981b). In other cases, such as *Gemelliporidra belikana* that lives in heavily grazed habitats on coral reefs, frontal budding always begins before the colony has reached ten zooids (M. Gleason & J. Jackson, unpublished data). Apparently, the increased relief of these heavily calcified colonies helps to resist grazing, for they are little affected.

The formation of erect foliaceous colonies from encrusting bases may be controlled by predation or other forms of disturbance. A good example of this is the sponge *Cliona celata* off North Carolina (Guida 1976). Under conditions of heavy grazing by sea urchins, this excavating sponge lives only in borings within calcareous substrata. In the absence of predators, however, it grows over the outer surface of its substratum and eventually forms erect branches. The same is probably true of western Atlantic *Steginoporella* spp. and *Stylopoma spongites* that usually grow only as encrusting colonies on coral reefs, but as foliaceous erect colonies offshore where there are fewer durophagous predators (J. Jackson, unpublished data).

All of the above changes in form can be explained as plastic developmental responses to changes in extrinsic environmental factors, including physical processes, food, competitors, and predators. These are the "unstable" species of Stach (1937). In contrast, most rigid, arborescent species are stable in that they begin life with encrusting growth, but very soon develop a more-or-less fixed vertical pattern, disrupted only by injury to the colony (Figs. 3.4 & 4.23, Cheetham & Thomsen 1981, Cheetham & Hayek 1983, Cheetham 1986a). The selective basis for this inflexible extra investment in support presumably involves benefits to the colony that accrue only in later stages of development (Jackson 1979a). Clearly, specialization in growth form decreases ecological options, thereby fixing the life history strategy of the species in many other ways. These ideas are pursued in the next chapter.

5 Bryozoan life histories

The life history of any organism consists of the sequence of developmental stages from birth to death, and the schedule of associated vital processes such as growth and sexual reproduction. Resources required for these processes are finite, and the demands of different environments vary, so that an organism's life history is, to some extent, a compromise. For example, a bryozoan that invests much energy into growing fast, or in building a colony of heavily calcified zooids, is unlikely to have as much energy left over to produce eggs as one that grows more slowly, or is only weakly calcified. Resources must be budgeted to potentially competing processes which can be broadly classified as growth, maintenance, and reproduction. In this way, a bryozoan's life history can be thought of as the consequence of a pattern of investment, or investment "strategy", which has been maintained and modified over evolutionary time by natural selection. Demonstrating which of these patterns of investment are associated with different conditions, and explaining their adaptive basis and origin, is one of the central problems of evolutionary biology.

The growth form model based on risk of mortality can also be used to make quantitative predictions about life history characteristics of bryozoans (Jackson 1979a, 1985, Jackson & Coates 1986). We have seen that, on average, uniserial or encrusting species are more frequent in relatively unstable environments. Accordingly, they should be opportunistic, so that they may propagate before the environment deteriorates. This implies a pattern of early maturity, high fecundity, simple or weakly calcified zooidal construction, and short life. In contrast, multiserial or erect species tend to predominate in relatively more stable environments, and should exhibit less opportunistic life histories, including late onset of reproduction, low or moderate fecundity, heavily calcified zooids, and long life.

In this chapter we first review some basic aspects of the reproductive ecology of bryozoans and describe the life histories of six species about which a great deal is known. We then discuss in a more general way the trade-offs that appear most important in bryozoan life histories, and relate these to predictions of the risk model. All this is of necessity purely biological. There are, however, important paleontological implications because much about the life history of bryozoan species is clearly preserved in their growth form and zooidal construction.

5.1 Reproductive ecology

All bryozoan colonies are hermaphroditic. Zooids usually produce both sperm and eggs, but cyclostomes and some ascophorans are dioecious,

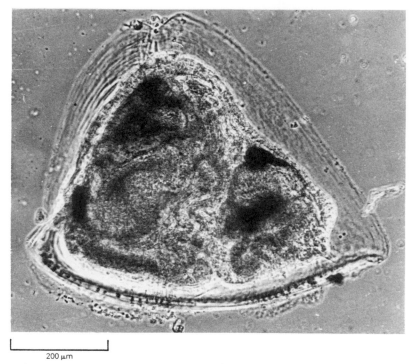

200 μm

Figure 5.1 Cyphonautes larva of *Membranipora serrilamella*. (From Mawatari 1975, courtesy of Nihon University School of Medicine.)

frequently with dimorphic male and female zooids (Fig. 9.10, Nielsen 1981). So far as is known, sperm are always released directly into the sea through pores in the tentacles (Silén 1966, 1972), but the fate of eggs is more variable. Two groups, anascans and ctenostomes, include a significant minority of species that release eggs directly into the sea, just before or after fertilization, where they develop to form free-swimming larvae (Zimmer & Woollacott 1977). The best known of these are the bivalved cyphonautes larvae (Fig. 5.1) of the anascans *Membranipora*, *Conopeum*, and *Electra*, which may swim for weeks in the plankton before settlement.

The most common pattern, however, is retention of the embryo somewhere on or within the maternal colony (Strom 1977). Among anascans, the larva may develop within the maternal autozooid whose polypide deteriorates to make room. This is also the pattern in all brooding ctenostomes. In most anascans, however, and almost all ascophorans, the embryo develops in special brood chambers termed ovicells. These tend to stick out above the colony surface in anascans, but are typically partially or completely buried by secondary calcification in ascophorans. Brooded gymnolaemate embryos develop into large larvae (Fig. 5.2) that are competent to settle more-or-less upon release. Cyclostomes brood embryos in a few large female **gonozooids**. Their embryogenesis is highly unusual

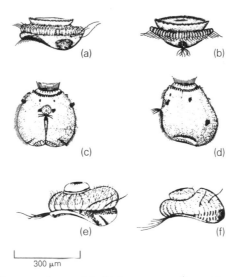

Figure 5.2 Brooded bryozoan larvae. (a), (b) Ctenostome *Alcyonidium mytili* (c), (d) anascan *Bugula simplex*; (e), (f) ascophorans *Escharoides coccinea* and *Cellepora pumicosa*. (From Ryland & Hayward 1977, Hayward & Ryland 1979 & 1985, courtesy of the Linnean Society of London.)

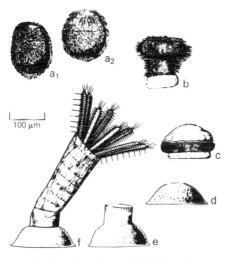

Figure 5.3 Larva metamorphosis and early growth of the cyclostome *Crisia eburnea*. (From Nielsen 1970, courtesy of *Ophelia*.)

(Harmer 1893). The fertilized egg goes through cleavage to form a primary embryo which then disassociates into separate blastomeres which in turn cleave to form secondary embryos. This process of **polyembryony** may be repeated so that the gonozooid becomes filled with large numbers of genetically identical embryos. Cyclostome larvae are smaller than brooded gymnolaemate larvae, but they, too, settle quickly.

Settlement involves cementation to a substratum, followed by meta-morphosis into an ancestrula (Fig. 5.3). The ancestrula commences budding new zooids rapidly, in some species before the ancestrular polypide has formed, suggesting that large larval energy reserves are available to begin colony development. There is also considerable variation in ancestrular budding patterns of cheilostomes (Fig. 5.4, Cook 1973). Some ancestrulae form as single zooids initially capable of budding in only one direction (Figs. 5.4a, b). Others, such as *Stylopoma spongites* (Fig. 5.4d), may metamorphose directly into complex ancestrulae of several zooids which may be capable of budding in all directions from the start. This variation is obviously related to colony form and survival in early life, but has not been investigated ecologically.

5.2   Six case studies

Life cycles of bryozoans are complicated by their **clonality**, which is the ability of parts of colonies to survive and reproduce on their own if separated from one another by injury or fission (Jackson 1977a, 1985, Hughes 1984, Jackson & Hughes 1985, Jackson & Coates 1986). **Aclonal** benthic animals like barnacles and mussels have simple life cycles (Fig. 5.5a). Numbers of individuals in a population increase only through recruitment of sexually produced larvae or by immigration, and decrease only through death or emigration. Individual animals are easily counted, and there is an approxi-mate genetically determined upper limit to body size. Most surviving individuals grow to about this upper limit, produce gametes or larvae, and eventually senesce and die. In contrast (Fig. 5.5b), numbers of colonies in a clonal population can increase by asexual as well as sexual reproduction, and decrease by fusion of previously separated colonies, as well as by death. Moreover, colony size can decrease considerably over time by fission or injury, stay constant, or increase by fusion or growth.

Very few bryozoans have been studied in sufficient detail to characterize adequately their life histories *in situ*. All but one of these are multiserial encrusting forms. There has been no comparable study of any rigidly erect species. It is already clear, however, that bryozoan life histories do not fall into two neat packages described by r- and K-selection as is, unfortunately, so often tried. Therefore, we describe each case in some detail to give a feeling for the natural variation involved.

5.2.1   Bugula neritina: *a true weed*

The cheilostome *B. neritina* is one of the most common and troublesome "fouling" organisms in temperate and subtropical seas, inhabiting almost any substratum in harbors around the world (Ryland 1965, Winston 1982). In the north-eastern Gulf of Mexico, however, it lives primarily on blades of

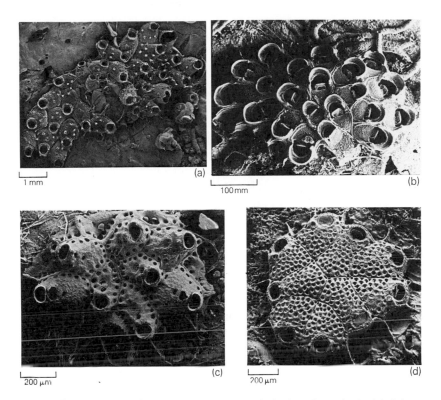

Figure 5.4 Early astogeny of cheilostomes. Asymmetrical colony formation by (a) *Coleopora americana* and (b) *Steginoporella magnilabris*, in contrast with the more radially symmetrical ancestrular budding of (c) *Reptadeonella* sp. and (d) *Stylopoma spongites*. In *Stylopoma*, the ancestrula forms as several zooids from the larva during metamorphosis.

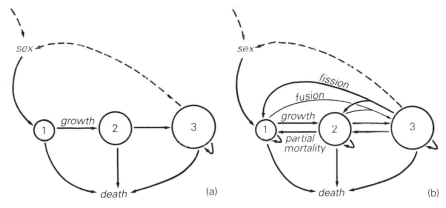

Figure 5.5 Schematic models of life cycles of benthic invertebrates; (a) aclonal, (b) clonal. (Redrawn from Jackson & Hughes 1985, courtesy of The Society of Sigma Xi.)

Figure 5.6 Underwater portrait of the anascan *Bugula neritina* growing on the seagrass *Thalassia* in the north-eastern Gulf of Mexico. The width of the *Bugula neritina* branches is approximately 300 μm. (Photograph courtesy of M. J. Keough.)

Thalassia and other seagrasses, where it is the most abundant bryozoan (Fig. 5.6, Keough 1986, Keough & Chernoff 1987), averaging one colony every 3–5 blades at peak season in early winter. Colonies are absent in summer when temperatures are high and blades grow as fast as 2 cm day^{-1}, or after midwinter freezes when colonies die. This is an highly unstable and ephemeral habitat. In addition to extreme variations in growth rate, old blades break off during winter or anytime during storms. Some colonies occur on tubes of the large worm *Diopatra* which do survive winter. However, both these habitats may be destroyed entirely when hurricanes rip out entire seagrass beds.

Bugula neritina colonies brood larvae that recruit in large numbers during the fall and winter. Besides this seasonality, recruitment varies greatly from year to year in relation to weather patterns. Larvae settle preferentially on the distal third of seagrass blades and on pieces of plastic ribbon used as experimental substrata.

Colonies grow as flexibly erect biserial bushes, up to about 8 cm tall. All branches bifurcate at approximately regular intervals, so that colony size can be estimated easily from the number of bifurcations along a branch. Growth is usually rapid but apparently very sensitive to temperature. For example, in one experiment, at 6 °C colonies took 30 days for the first bifurcation, and in another at 21 °C, only six days. Zooids are weakly calcified and the ovicells stick out well above the surface of the branch. Unlike most other species of *Bugula*, there are no avicularia or spines.

Some colonies produce embryos after the third bifurcation, and almost all do after 5–7 bifurcations, when most zooids have ovicells containing embryos. This translates to between <2 and >9 weeks for reproduction to begin. Survival is very short. About 70% of all recruits on plastic ribbons died within one week, and after 2–3 weeks only a few percent were left, all on the distal portions of these "blades". After this time, survival increased greatly. Colonies on seagrass blades cannot live more than six or seven months, but those on *Diopatra* may "oversummer" by dying back and resuming growth and reproduction in the autumn. Thus they are potentially perennial.

In summary, *B. neritina* is a uniserial, morphologically simple, opportunistic species living in an highly disturbed environment. This pattern is as predicted by the risk model.

5.2.2 Membranipora membranacea: *a specialized weed*

Along the coast of California, *M. membranacea* lives primarily on blades of the giant kelp *Macrocystis pyrifera* (Fig. 5.7, Woollacott & North 1971, Bernstein & Jung 1979, Yoshioka 1982a). As for seagrass, this is an highly ephemeral habitat because the kelp blades live no longer than six months (North 1961), and the abundance of kelp varies greatly due to storms,

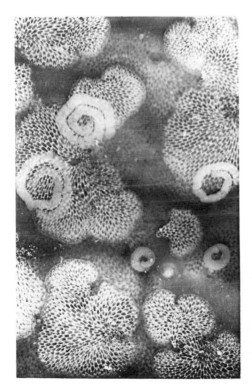

Figure 5.7 Underwater photograph of young colonies of the anascan *Membranipora membranacea* growing on the kelp *Macrocystis pyrifera* near San Diego, California. Note the spiral egg masses of nudibranchs that prey upon *Membranipora*. The length of the *Membranipora membranacea* zooids is approximately 700 μm. (Photograph courtesy of P. M. Yoshioka.)

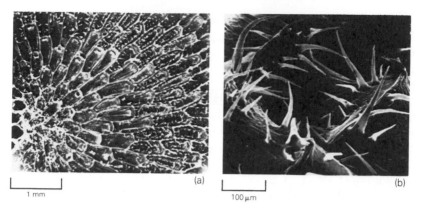

(a)

1 mm

(b)

100 μm

Figure 5.8 Spine induction in *Membranipora membranacea*. (From Harvell 1984a, courtesy of the American Association for the Advancement of Science.)

herbivores, and the life cycle of the plants (Dayton *et al*. 1984). The cyphonautes larvae live 2–4 weeks in the plankton and are thus inevitably dispersed far from the parental colony. The larvae settle preferentially on *Macrocystis* when swimming at the surface, but on other algae near the bottom when warm temperatures restrict them to deeper water. Recruitment onto kelp and numbers of colonies per blade are extremely variable from season to season and year to year, a common feature of animals with long-lived planktonic larvae (Thorson 1950, Caffey 1985).

Colonies are usually small because of crowding at settlement. Nevertheless, growth is indeterminate, as shown by occasionally very large colonies on uncrowded fronds. Growth is extremely rapid, as much as 2 mm day^{-1}, or roughly 6 cm month^{-1}. With such rapid growth, *M. membranacea* colonies easily overgrow all other epibionts they encounter. Zooids are only weakly calcified, which may favor survival on their flexible host. Consequently, they are also vulnerable to biting or crushing predators like fish, some of which feed preferentially on kelp that is heavily encrusted by the bryozoan (Bonsdorff & Vahl 1982). Two nudibranchs mimic *M. membranacea* and feed exclusively upon it, but their feeding may be slowed dramatically by spines whose formation is induced by the presence of the nudibranchs (Fig. 5.8, Yoshioka 1982b, Harvell 1984a).

The colonies reach sexual maturity within a few days, when only 5 mm diameter. These colonies produce as many eggs per zooid as much larger colonies. Reproduction continues throughout the life of the colony, and fecundity is enormous. The vast majority of colonies die within a few months with death of kelp blades, but those on other substrata might survive much longer.

In summary, *M. membranacea* is highly opportunistic, matching well most of the predictions of the risk model. However, it is also highly specialized for life on kelp blades, as evidenced by larval habitat selection, a flexible skeleton, and even the uptake of dissolved organic material exuded by kelps

(DeBurgh & Fankboner 1979). *Membranipora* dominates in this habitat due primarily to massive recruitment and rapid growth, and also because no other sessile organisms successfully oppose it.

5.2.2   Steginoporella *sp. and* Reptadeonella costulata: *mobile and stationary perennials*

These two cheilostomes encrust the undersurfaces of foliaceous reef corals in Jamaica, where they typically cover more space than any other bryozoans

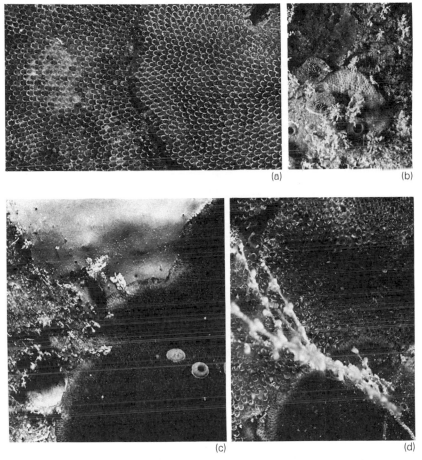

Figure 5.9   Underwater photographs of bryozoans encrusting undersurfaces of foliaceous reef corals, Rio Bueno, Jamaica. (a) *Steginoporella* sp. at left growing over older region of the same colony; (b) *Drepanophora tuberculatum* surrounded to the left and below by three osculae of a boring sponge growing within the coral substratum and by crustose algae to the top and right; (c) *Reptadeonella costulata* growing over and being overgrown by the foraminiferan *Gypsina* (top) and overgrowing various algae (bottom); (d) *Reptadeonella* (below) overgrowing old zooids of *Steginoporella* from behind, with a stalked hydroid growing above them. The length of *Steginoporella* zooids and diameter of sponge osculae are approximately 1 mm.

(Fig. 5.9, Jackson & Buss 1975, Jackson 1979b, 1984, Jackson & Winston 1982, Winston & Jackson 1984, Jackson & Hughes 1985). The corals which provide their substratum are long lived, the majority persisting for decades and many for centuries (Hughes & Jackson 1985). Their undersurfaces are sites of intense competition for space, where growth of any creature usually involves partial or complete death of another.

Both species brood larvae that settle in very small numbers throughout the year; recruitment is at least three orders of magnitude less than for *M. membranacea*. Ancestrulae of *Steginoporella* are not uncommon under corals, but those of *R. costulata* are rare. Habitat selection has not been studied; *Steginoporella* sp. lives only on reefs, but *R. costulata* occurs commonly in some other habitats.

Colonies of both species grow to be large, commonly reaching 100–200 cm^2 or more (Jackson & Wertheimer 1985). *Steginoporella* grows as lobate fans in shallow water, but may also form narrow ribbons at greater depths (Fig. 5.10). This pattern is intermediate between uniserial and multiserial growth. The leading edges of the fans advance across the coral as fast as 11 cm year^{-1}, rarely remaining in any one place for more than six months. Zooids are morphologically simple, with little secondary calcification, an enormous orifice, and thin frontal and lateral walls (Fig. 5.11a, b). The only polymorphs, termed "B" zooids, are similar in appearance to autozooids except for typically greater size and calcification; these may be a primitive form of avicularium. In contrast, *R. costulata* forms roughly symmetrical colonies that grow about 3–4 cm year^{-1} and commonly persist in any one spot for several years. Zooids are complex, with obvious surface fortifications, and thick frontal and lateral walls (Fig. 5.11c, d). There are also adventitious avicularia and female gonozooids. Paralleling these differences, *R. costulata* zooids are about 15 times harder to puncture, and colony surfaces nearly twice as hard to crush, as are those of *Steginoporella* (Best & Winston 1984).

Both species require about two years before they begin to produce embryos (Winston & Jackson 1984). At that time, *R. costulata* may be one-third the size of *Steginoporella* sp., and may have about four times more embryos per cm^2 of colony surface (Jackson & Wertheimer 1985). Standing crop of embryos increases as a linear function of colony size in both species (Fig. 5.12). Although large colonies of both species may live for years, median survival of newly recruited *Steginoporella* onto settlement panels was only 120 days, whereas *R. costulata* survived for about 500 days.

We can better interpret these differences in life histories by comparing the functional biology of the two species (Jackson 1979b, Palumbi & Jackson 1982, 1983). *Steginoporella* colonies display marked gradients between younger and older regions (Fig. 5.9a), correlated with senescence of older zooids, whereas *R. costulata* shows no obvious regional senescence of differentiation of function. Younger regions of *Steginoporella* colonies are brightly colored, unfouled by other organisms, and unbroken; older regions

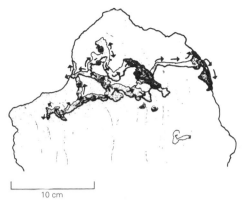

10 cm

Figure 5.10 Drawing of a highly mobile, ribbon-like colony of *Steginoporella* sp. from under a coral collected from 60 m depth, Discovery Bay, Jamaica. The pentagon-shaped formation indicates the approximate point of larval settlement (ancestrula eroded away) and the arrows the directions of colony growth. The two areas alive when collected are indicated by darker shading; these comprise about one-sixth of the total colony area. (From Jackson & Winston 1981, courtesy of Olsen & Olsen.)

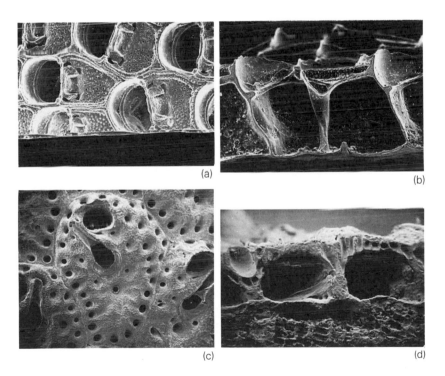

(a)

(b)

(c)

(d)

Figure 5.11 Zooidal architecture of abundant cheilostomes under Jamaican corals. (a) *Steginoporella* surface (zooids ~ 1.1 mm long) and (b) cross section; (c) *Reptadeonella costulata* surface (zooids ~ 0.7 mm long) and (d) cross section. (From Jackson & Hughes 1985, courtesy of The Society of Sigma Xi.)

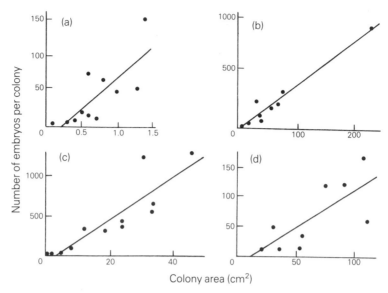

Figure 5.12 Relationship between numbers of embryos per colony and colony area for four encrusting species living under corals near Discovery Bay, Jamaica: (a) *Drepanophora tuberculatum*; (b) *Reptadeonella costulata*; (c) *Parasmittina* sp.; (d) *Steginoporella* sp. (From Jackson & Wertheimer 1985, courtesy of Olsen & Olsen.)

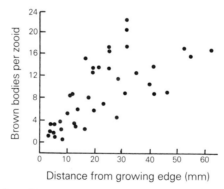

Figure 5.13 The number of brown bodies per zooid at different distances from the growing edge of colonies of *Steginoporella* sp. (From Palumbi & Jackson 1983, courtesy of *Biological Bulletin*.)

are dark, fouled, and broken, These differences coincide with the progressive accumulation of brown bodies in *Steginoporella* zooids, which do not accumulate in the other species (Fig. 5.13).

The deterioration of older regions of *Steginoporella* colonies affects its ability to function in at least three important ways. Firstly, older regions are much poorer overgrowth competitors than younger regions (Table 5.1). Second, older regions feed less frequently than younger zooids of the same colony (Fig. 5.14). Third, rates of regeneration of injuries are less than one-fourth as fast in older compared to younger regions (Fig. 5.15). No such variation is apparent in *R. costulata* which regenerates injuries at rates almost equal to young *Steginoporella*.

Steginoporella colonies survive through a balance of spatially partitioned growth and decay. Rapid growth of younger regions provides exceptional competitive and regenerative ability, but at the expense of localized

Table 5.1 Outcome of interspecific overgrowth interactions involving younger (distal) and older (proximal) areas of *Steginoporella* sp. colonies.

Colony	Interactions		
Region	wins	losses	total
distal	34	5	39
proximal	2	20	22

2×2 contingency table, $\chi^2 = 35.5$, $P < 0.001$.

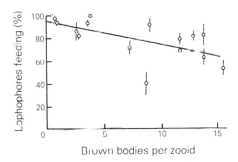

Figure 5.14 Percentage of zooids with lophophores feeding in different parts of *Steginoporella* sp. colonies as a function of the numbers of brown bodies per zooid in the different areas. (From Palumbi & Jackson 1983, courtesy of *Biological Bulletin*.)

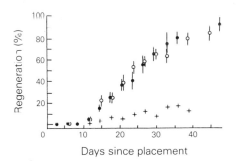

Figure 5.15 Rates of regeneration of 4 mm diameter lesions by *in situ* colonies of *Steginoporella* sp. and *Reptadeonella costulata*. Open circles, *R. costulata*; closed circles, young areas of *Steginoporella* near the growing edge; crosses, older areas of *Steginoporella* away from the edge. (From Palumbi & Jackson 1982, courtesy of Elsevier Biomedical Press.)

senescence of zooids that prevents them from holding on to any patch of substratum. *Steginoporella* apparently invests so much in growth that it has few resources left for reproduction until colonies become large, whereas *R. costulata* invests more in fortification and maintenance of zooids, and in sexual reproduction. Circumstantial evidence, particularly the marked polarity of regeneration in *Steginoporella* colonies (Fig. 5.16), suggests that these patterns result from variations in the extent and direction of translocation of resources between zooids.

These differences in physiology and life history are reflected in the contrasting distributions of the two species as a function of distance from the edges of their host corals (Jackson & Hughes 1985). The edge is where space is produced by coral growth, and where grazing by sea urchins and fishes is most intense. It is also where sponges, which are the best overgrowth competitors under corals, are least abundant. *Steginoporella* is relatively more abundant nearer coral edges. As predicted by the risk model, rapid growth is apparently more adaptive than strong skeletons in the more disturbed environment near coral edges.

5.2.3   Drepanophora tuberculatum *and* Disporella fimbriata: "solitary" colonies

The cheilostome *D. tuberculatum* and cyclostome *D. fimbriata* are very

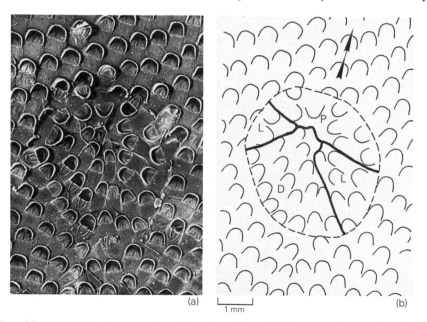

(a)

1 mm

(b)

Figure 5.16   Polarity of regeneration into a 4 mm diameter lesion near the growing edge of *Steginoporella* sp.; (a) photograph, (b) schematic diagram showing amounts of regeneration in distal (D), lateral (L), and proximal (P) directions. (From Palumbi & Jackson 1983, courtesy of *Biological Bulletin*.)

common under foliaceous corals in Jamaica (Figs. 5.9c & 5.17, Jackson 1983, Winston & Jackson 1984, Winston 1985, unpublished data). In addition, *D. tuberculatum* is the most common cheilostome under small coral rubble in shallow water, substrata that are destroyed by excavating organisms, grazers, and storms within a few years. In contrast, *D. fimbriata* is also common in reef tunnels and caves.

Both species brood larvae that recruit onto settlement panels and natural substrata at rates at least ten times greater than for *R. costulata* and *Steginoporella* combined. Numbers of recruits vary markedly in space and time, but there is no clear seasonal pattern. There is some evidence for habitat selection: recruits of both species are more common on panels placed in their primary habitats.

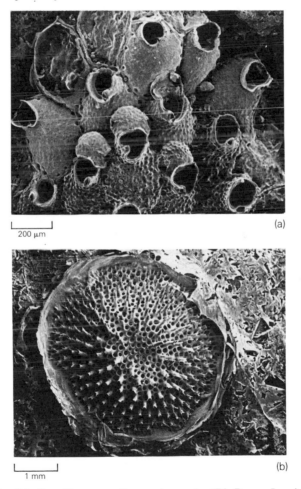

200 μm   (a)

1 mm   (b)

Figure 5.17 Small "solitary" bryozoans from under corals at Rio Bueno, Jamaica: (a) half of a colony of *Drepanophora tuberculatum* showing ovicells present at 1–2 zooids' distance from the colony margin (note also extensive secondary calcification of zooids), (b) *Disporella fimbriata.*

Upon settlement, *D. tuberculatum* grows rapidly, reaching a maximum colony size of 2–3 cm^2 within six months or less. Zooids are heavily calcified, and colonies are more resistant to puncture or crushing than any other species tested from Jamaican reefs (Best & Winston 1984). Ovicells are sunken beneath extensive secondary calcification. Grazing scars are common around colonies, but the colonies themselves are not usually badly damaged and can regenerate. However, they are commonly overgrown by other bryozoans and sponges. *Disporella fimbriata* also grows rapidly to a maximum size of about 1 cm^2 within 3–6 months, forming a discoidal dome with a raised margin that helps prevent overgrowth, and a single large central brood chamber exposed at the surface. Even partially grazed colonies do not usually survive.

Colonies of *D. tuberculatum* release larvae within three months, when less than 3 mm in diameter, and continue to reproduce until about six months old. Most zooids have ovicells that contain embryos throughout this period; embryos are produced and released by individual zooids on an approximately weekly cycle. After six months, most colonies die or become fouled, with scattered senescent patches that divide the colony into separate subcolonies. Some of these survive to undergo one or more cycles of regrowth and reproduction, but with diminished vigor and size. Median survival is about 230 days. *Disporella fimbriata* also develop larvae within three months. Afterwards, they usually die, but some may survive to reproduce a few months later, and a very few may survive for three years. Symptoms of senescence include breakdown of the colony margin and dissolution of the zooidal tubes.

These species illustrate two ways that bryozoans can be successful by being small and "solitary". For *D. tuberculatum*, semi-determinate growth allows both fortification and prolific reproduction, while periodic regression into subcolonies that may rejuvenate provides greater opportunity for continued survival. *Disporella ovoidea* is even more like a solitary animal that grows, invests little in defense, reproduces, and dies.

5.3 Life history patterns

Trade-offs between competing demographic processes have resulted in several life history patterns that are correlated with growth form of colonies.

The limited data available suggest that relationships between recruitment, growth, and mortality are consistent with predictions of the growth form model. Among encrusting species on Caribbean reefs, for example, uniserial *Aetea* grows up to 6 cm month^{-1} (72 cm year^{-1}) compared with 3–11 cm year^{-1} for *R. costulata* and *Steginoporella* sp., and only 1 cm year^{-1} for mound-like *Trematooecia aviculifera* (Jackson & Hughes 1985, M. Gleason unpublished data, J. Jackson & K. Kaufmann unpublished data). Similarly, 590 recruits of *Aetea* settled onto experimental panels during one month, compared with only five recruits of the three multiserial species combined. In contrast, the extent of calcification of zooids and colony-wide regenerative ability show the opposite pattern.

The ratio of percentage cover of uniserial to multiserial species is extremely low under corals but high for numbers of colonies recruiting onto settlement panels put out for 3–4 months in the same environment (Table 5.2). Moreover, the ratio for percentage cover of encrusting to erect species is extremely high under corals, but very low for numbers of recruits. There are no such differences, however, for numbers of species instead of colonies or cover. Uniserial species were common on panels because they recruit in much higher numbers, and had not yet been eaten or pushed over and overgrown (Buss 1981a). These patterns parallel those for distributions of growth forms with stability of substrata and salinity (§ 4.4.3–4).

Differences in reproductive effort between growth forms are undoubtedly even greater than the data suggest, because most gametes or larvae die before settling, and because settlement panels are usually left in the sea so long (≥ 1 month) that most of the earliest settlers are gone by the time recruits are counted. For example, Grosberg (1981) measured recruitment daily in the Eel Pond at Woods Hole, and Osman (1977) made monthly observations at a neighboring island < 2 km away. Panels were suspended 1 m below a dock in both studies. The uniserial to multiserial ratio for species of bryozoan recruits measured daily was 1.0, versus only 0.25 from monthly observations. Ratios of encrusting to erect species were 1.7 measured daily, and 4.0 measured monthly. The more frequent and therefore accurate the observations, the more the results are in accord with predictions of the risk model.

There may also be a correlation between growth form and longevity in bryozoans. Species were sorted into three groups based on reported lifespans: < 1 year, annuals or biannuals, and perennials (Eggleston 1972b). Some of the data are suspect, because colonies of several species may die back during winter but regenerate in spring (e.g., *B. neritina*). Nevertheless, the pattern is striking. Ratios of numbers of uniserial to multiserial species

Table 5.2 Differences between bryozoans under corals and those on settlement panels after 3–4 months at 10 and 20 m depths on north-coast Jamaican reefs (Jackson 1984, J. Winston & J. Jackson unpublished data).

Ratio	Percentage cover under corals	Number of colonies on panels	Number of species	
			under corals	on panels
uniserial/ multiserial	0.03	3.9	0.23	0.24
encrusting/ erect	75	0.25	4.0	4.2

(0.39), and of encrusting to erect (4.1), are much higher for species living for <2 years than for perennials (ratios 0.09, 0.50). This is consistent with predictions of the risk model.

Many bryozoans reproduce as much or more by asexual as by sexual means. Asexual reproduction may occur by localized partial mortality of zooids, by fragmentation of colonies, and by fission or budding. Partial mortality is the most common process among encrusting species; more so among highly mobile uniserial or fan-shaped colonies such as *Stomatopora* sp. and *Steginoporella* sp. (Figs. 4.4, 5.10) than among more symmetrical, stationary species such as *R. costulata*. Fragmentation is most frequent among rigidly erect and free-living species, pieces of which commonly regenerate into entire new colonies (Fig. 9.13, Winston 1983, 1986). Budding occurs in some free-living colonies on sediments (Fig. 5.18, § 9.1.2). All of these processes are readily distinguishable in fossils (Chs. 8, 9, Jackson 1983).

Erect species that regenerate frequently from fragments commonly show reduced rates of larval recruitment, as evidenced by small numbers of colonies originated from ancestrulae (Winston 1983). This pattern is also characteristic of erect fossil species (§ 8.3.1–2) as well as for living and fossil free-living species (§ 9.1.2).

5.4   Size and age

Because of differential rates of positive and negative growth among colonies, the schedule of life history events in bryozoans is closely tied to their size as well as to their age (Jackson 1985, Jackson & Hughes 1985, Keough 1986, Keough & Chernoff 1986). Ability to compete for space, resist predators, regenerate injuries, or reproduce sexually all increase with colony size. A large colony that has become small due to injury is more likely

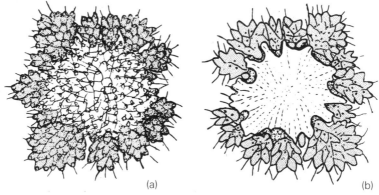

Figure 5.18 Zoarial budding of colonies by *Discoporella umbellata depressa* from Brazil. (Redrawn from Marcus & Marcus 1962.)

to die and less likely to reproduce than when it was larger, and may thus more closely resemble other small colonies demographically, whatever their age, than its contemporaries that remained large.

Production of bryozoan larvae per zooid, or per unit area of colony surface, is roughly constant in most species for colonies above some minimum colony size (Fig. 5.12, Hayward & Ryland 1975), so that total colony production should increase proportionally or exponentially with colony size. Since few colonies of potentially large species appear to die of old age (Palumbi & Jackson 1983), and chances of survival and sexual reproduction both increase with colony size, some bryozoan populations may become dominated by relatively few long-lived, highly fecund clones. In such cases, the evolutionary stakes for achieving large colony size are clearly enormous.

5.5   Dispersal

The great majority of bryozoans brood short-lived larvae ready to settle within seconds to hours of release by their parents (Ryland 1981, Jackson 1985, 1986). Thus it is extremely unlikely that brooded larvae could be transported more than a very short distance (centimeters to a few kilometers). Indeed, experiments show that many, perhaps most, larvae of *Lichenopora* living on kelp may travel only a few centimeters before settlement (C. MacFadden, unpublished data). Moreover, frequent fusion of young colonies of Paleozoic stenolaemates suggests a long history of brief larval travel (Palmer & Palmer 1977, McKinney 1981a). Such short-distance dispersal has profound implications for patterns of distribution and evolution of bryozoan species.

5.5.1   Patterns of distribution

Populations of *Bugula neritina* are extremely patchy between apparently similar seagrass beds separated by only a few kilometers or less (Keough & Chernoff 1986). Transplant experiments show that these patterns are not due to subtle environmental differences between sites, because colonies grow and reproduce as well or better in beds where *B. neritina* is naturally absent. The most likely explanation for this patchiness is very low dispersal between beds. Similarly, disjunct local distributions of reef-associated bryozoans are characteristic throughout the Caribbean (J. Winston & J. Jackson, unpublished data).

How then are bryozoans dispersed, especially to oceanic islands like Hawaii? The most likely mechanism is rafting on algae, wood, pumice, or other natural floating debris (Jackson 1986), or in historical times on ships and human garbage (Ryland 1965, Winston 1982). In the case of *B. neritina* in the north-eastern Gulf of Mexico, colonies commonly occur on floating

Thalassia blades that have been torn from their shoots (Keough & Chernoff 1986). Such bryozoans are healthy and often contain embryos. Seagrass blades may float for considerable distances before washing ashore. In doing so, they often pass over seagrass meadows where fortuitously released larvae could settle. For many such species, rafting is probably the only means of dispersal between sites. In this case, the primary value of larvae for dispersal should be for getting onto and off rafts.

A striking consequence of rafting is the lack of association between length of larval life and species' distributions, even between brooding species and those with long-lived planktonic larvae (Jackson 1986). This conclusion is based on analysis of four western Atlantic faunas from the Caribbean and eastern Florida (Winston 1982, Jackson *et al.* 1985). Species were scored as occurring in none, one, two, or three other tropical regions around the world. The faunas include 65 reef-associated species, all brooders, and seven species with cyphonautes larvae (Table 5.3). The geographic distributions of these two groups are virtually identical.

Table 5.3 Comparison of geographic distributions of tropical and subtropical western Atlantic cheilostomes with cyphonautes versus brooded larval development. Data are numbers of western Atlantic species also found in some combination of three other tropical to subtropical regions: (1) eastern Atlantic and Mediterranean, (2) western Pacific to the Red Sea, and (3) eastern Pacific from Hawaii to the Americas.

Larval mode	Number of other regions inhabited			
	0	1	2	3
cyphonautes	2	2	1	2
brooded	25	16	13	11

2×4 contingency table, $\chi^2 = 0.73$, $P > 0.8$.

5.5.2   Evolutionary consequences

Frequent asexual reproduction coupled with short-distance dispersal of brooded larvae should increase genetic relatedness in local bryozoan populations and decrease gene flow between them. This should, in turn, increase chances of evolutionary adaptation to local environmental conditions and of speciation. Frequent rafting, however, would tend to counteract these processes.

Genetic and morphologic differentiation suggesting little gene flow has been observed for the cheilostome *Schizoporella errata* along Cape Cod (Schopf & Dutton 1976). Statistically significant changes occur over distances as little as 11–13 km (Fig. 2.4). However, despite this local differentiation, between 80 and 89% of the electrophoretically sampled genome of this species is identical along 1,000 km of the eastern coast of North America. We interpret these patterns to mean that short larval dispersal accounts for local population differences, but rafting is sufficiently frequent along the coast to prevent regional differentiation. In contrast, *B. neritina* from southern California and Florida, long isolated from each other by the uplift of the Panamanian Isthmus, show marked differences in juvenile growth rates and size at first reproduction (M. J. Keough, personal communication). These differences are maintained when colonies from the two populations are grown in a common laboratory environment.

Bryozoan species are eurytopic, occurring on a wider range of substrata, and over a wider depth range, than many other sessile animals, such as reef-building corals (Jackson *et al.* 1985). This suggests that bryozoans should also tend to be widely distributed geographically (Jackson 1974, Jablonski & Valentine 1981). Indeed, reef-associated cheilostomes are much more widely distributed than the corals they often live on, and the number of habitat types occupied by a species is significantly positively correlated with its geographic range (Jackson *et al.* 1985). These patterns are not due to differences in larval dispersal, because all the bryozoans involved brood short-lived larvae, while only some of the corals do.

Another consequence of eurytopy and broad geographic range should be low probabilities of both speciation and extinction, and thus long species durations (Jackson 1974). Comparison with the history of reef corals on opposite sides of the Panamanian Isthmus supports this prediction in a general way (Jackson *et al.* 1985), but there are too few rigorous evolutionary studies to say anything more (Ch. 2).

5.6   Heterogeneity within colonies

As already seen for *Steginoporella* sp., the structure, condition, and function of different parts of bryozoan colonies may vary widely, even in simple encrusting forms lacking any need for localized structural support (Boardman & Cheetham 1973, Ryland 1979b). Variation may be due to intrinsic developmental and physiological gradients within colonies, or to extrinsic events. Zooidal morphology and size, lophophore morphology, accumulation of brown bodies, production of gametes or embryos, onset of frontal budding, extent of secondary calcification, and occurrence of spines may all change along simple or complex proximal to distal gradients, or in more haphazard fashion (Correa 1948, Banta 1972, Dudley 1973, Stebbing 1973b, Dyrynda 1981, Dyrynda & Ryland 1982, Harvell 1984a, b, Jackson & Wertheimer 1985, McKinney *et al.*, 1986a). Moreover, in all species tested, these gradients are accompanied by additional physiological differences in feeding behavior, regenerative ability, or senescence (Ryland 1979b, Palumbi & Jackson 1983, Harvell 1984b).

The relationships between gradients within colonies and life histories of bryozoans are variable and complex (Jackson 1979a). For example, accumulation of brown bodies and increased production of embryos in older, more proximal areas of colonies of *Dendrobeania lichenoides* is associated with decreased nudibranch predation relative to that in more distal, younger areas (Harvell 1984b). In contrast, grazing on *Steginoporella* sp. by sea urchins and fishes occurs primarily on older zooids (Jackson & Kaufmann 1987) where there are more brown bodies, but not more embryos (Jackson & Wertheimer 1985). Presumably these grazers prefer the older regions because of the abundance of epiphytic algae there, rather than the accumulated brown bodies. Regardless, the relation of predation to accumulation of brown bodies is opposite in the two species.

The considerable variation in the nature and pattern of gradients within different bryozoans, along with their ecological importance, suggests that they are not simply due to physiological constraints of clonal growth, and may therefore represent specific ecological adaptations. Evolutionary interpretation is extremely difficult, however, because different parameters of growth, maintenance, and reproduction are almost certainly interdependent. Thus, the life history of an encrusting bryozoan that refrains from sexual reproduction until it has occupied all the substratum available might reflect past selection for rapid pre-emption of space. Alternatively, or perhaps concurrently, onset of reproduction when there is no place left to grow may simply mean that reproduction in the particular species can only occur when resources are no longer being invested in growth. The potential for circular reasoning in such matters is obvious.

6 Feeding: a major sculptor of bryozoan form

Bryozoans are suspension feeders that extract food from currents generated by their zooidal feeding structures. The sizes, and therefore types, of food captured by any species depend on the sizes of the individual feeding structures and their feeding behavior. Feeding behavior varies widely but predictably with colony shape and with the proximity and degree of functional integration between individual zooids in the colony. Fortunately, all of these features correlate well with zooid and orifice size, thereby allowing inference of diet from readily measurable skeletal features of both living and fossil forms.

The necessity of expelling filtered water so that it is separated from incurrent water has placed limits on the width of most branch types throughout the history of the phylum, and requires that broad surface areas either have relatively few zooids feeding at one time or be organized into cooperative feeding groups centred around a spot where filtered water is rapidly expelled. Relative constancy in sizes and spacing of zooids and branches and in subcolony or colony size throughout bryozoan history strongly suggest that bryozoans have had similar feeding patterns throughout their existence and have fed on the same range of plankton sizes.

6.1 Food

The primary food of shallow water bryozoans is unarmored phytoplankton, less than 50 μm in diameter. However, species of *Zoobotryon*, *Bowerbankia*, and others that have a gizzard can grind and eat diatoms, and at least one species eats zooplankton. Some bacteria are probably taken in during feeding, but they are, in general, too small to be captured by bryozoans (Winston 1977b), and the few laboratory data available indicate that bacteria are insufficient as food for bryozoans.

Not all suitably sized phytoplankton are sufficient foods, however. Some are toxic, while others differ strikingly in food value for different bryozoans, even within a single genus (Jebram 1975, Winston 1976). Quality of food has a pronounced effect on size, shape, and structure that colonies attain in laboratory cultures (Winston 1976, Kitamura & Hirayama 1984). Colonies of the encrusting species *Conopeum tenuissimum* are larger in area and are more nearly circular when grown 'on good food, are smaller and more indented in outline when grown on fair food, and show paltry growth with few functioning polypides when only poor food is provided (Fig. 6.1,

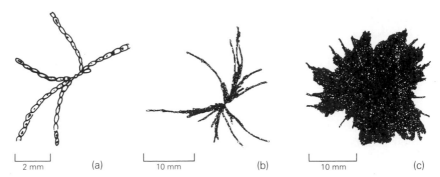

|___| (a) |___| (b) |___| (c)
2 mm   10 mm   10 mm

Figure 6.1   Variation in colony form of *Conopeum tenuissimum* after about 40 days' growth in the laboratory, grown on (a) the chlorophyte *Dunaliella tertiolecta*, (b) the chrysophyte *Monochrysis lutheri*, and (c) the diatom *Cyclotella nana*. (From Winston 1976, courtesy of the Marine Biological Laboratory, Woods Hole.)

Winston 1976). Such morphological differences should be visible in the fossil record where, for example, the same species lived in nutrient-poor lagoons and relatively nutrient-rich open waters. Two warnings for paleontologists arise from laboratory studies on effect of diet on bryozoan colonies:

(a) colony shape can be affected as much by interaction with the biological environment as by physical causes, and
(b) colonies of the same species growing in different environments, or at different times of the year (when different food types were available) within the same environment, may look radically different.

The only available field data on feeding suggest that sponges compete with bryozoans for unarmored phytoplankton (Buss & Jackson 1981). Absence of bryozoans from an otherwise suitable habitat may be due to depletion of their food source by sponges as well as to simple overgrowth competition (Ch. 7).

Phytoplankton are not the only food of all bryozoans. Those with large tentacle bells may supplement or replace a phytoplankton diet by capturing ciliates or other mobile protistans. *Bugula neritina*, for example, forms cages around such food particles by pulling the tips of tentacles in a lophophore together, and then ingests the prey (Winston 1978). Bryozoans that live in deep water must have some food source other than live phytoplankton. Their nutrition may come from detritus, bacteria, or nutrients absorbed directly. An examination by Schopf (1969) of guts of deep-water bryozoans found them to be empty after the 1.5–2 hour travel time from sea floor to ship. However, fecal pellets of these bryozoans contain calcite, indicating ingestion if not digestion of detritus.

Various behaviors, including flicking of tentacles and retraction of lophophores, help to prevent ingestion of unwanted particles (Winston 1978). Sediments, fecal pellets, and other bits of trash are flushed away from colony surfaces by filtered water that flows to ejection regions such as chimneys, colony margins, or slots between adjacent thin branches (Cook 1977a, Winston 1978, Lidgard 1981). Cook (1977a) reported that a 5 mm diameter *Lichenopora* colony cleared itself of a thin layer of fine mud within 10–15 minutes.

6.2 Feeding structures

6.2.1 Lophophores and mouths

Bryozoans feed from currents generated by ciliated tentacles that extend from a ring-shaped base around the mouth of each individual polypide. In addition to driving the food-bearing current, the cilia of the tentacles trap food particles, which are typically stopped and passed frontally by the lateral cilia and travel to the mouth along the frontal cilia (Atkins 1932, Strathmann 1973, 1982, Ryland 1976, Best & Thorpe 1983).

Growing edges or tips of colonies typically lack functional polypides in the youngest developing zooids. Behind this zone in cyclostomes there is commonly a zone of ontogenetic increase in polypide and tentacle size where tentacle length increases after polypides have begun to feed (Fig. 6.2); polypides of young zooids are large in cheilostomes. Tentacles vary in average length and number from species to species, and both are on average less in cyclostomes than in ctenostomes and cheilostomes (Table 6.1).

Tentacles are spread into various funnel- or bell-shapes when the zooids are protruded to the feeding position. Individual lophophores may be

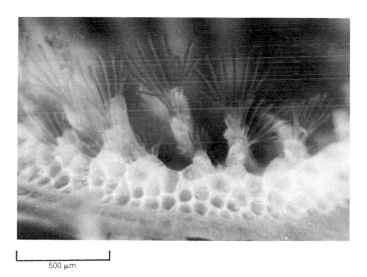

500 μm

Figure 6.2 Colony margin of the cyclostome *Plagioecia patina*, showing ontogenetic increase in size of tentacles from colony edge to zone where peristomes are well developed. (From Silén & Harmelin 1974, courtesy of The Royal Swedish Academy of Sciences.)

Table 6.1 Sizes of feeding structures in Recent marine bryozoans. Data are from Ryland (1975) and Winston (1977b, 1978, 1979). Significant differences ($P < 0.05$) exist between stenolaemates and each of the other three groups for each parameter (Fisher's LSD multiple comparisons).

	Stenolaemata	Ctenostomata	Cheilostomata	
			Anasca	Ascophora
Tentacle number				
mean	9.6	14.0	15.2	15.1
standard deviation	2.5	7.8	4.4	3.5
range	8–16	8–31	9–26	11–23
number of species	14	20	46	44
1–way ANOVA; $F = 5.78$, $P = 0.001$				
Tentacle length (µm)				
mean	317	455	456	513
standard deviation	155	242	201	178
range	166–698	173–1010	210–1094	167–929
number of species	14	20	46	43
1-way ANOVA; $F\ 3.52$, $P = 0.02$				
Mouth diameter (µm)				
mean	22	39	39	42
standard deviation	8	27	16	17
range	15–42	18–102	17–91	16–86
number of species	10	16	31	28
1-way ANOVA; $F\ 3.00$, $P = 0.03$				

composed of tentacles equal in length, or which grade from long on the anal side to short on the other. Filtered water is dispersed radially below equi-tentacled lophophores. Graded tentacle lengths result in obliquely truncate lophophores, below which filtered water flows away from the zooid on the side of the longer tentacles. Lophophore shape is closely related to colony shape, and thus can be estimated for fossil bryozoans of known colony morphology (§ 6.3, 6.4).

Size of food particles that can be ingested is limited by mouth size (Fig. 6.3). The observed range of mouth sizes is 15–102 µm, has low variability within a species (Winston 1977b, 1978, 1981), and averages less for cyclostomes than for ctenostomes and cheilostomes. Cyclostomes as a group therefore cannot feed on as broad range of food particles as can ctenostomes and cheilostomes. Among the reasons for the Cretaceous to Recent increase in cheilostomes over cyclostomes may be displacement of stenolaemates from the upper range in size of their potential food sources. Mouths less than about 30 µm are all circular, whereas those in the upper part of the size range are elongated, generally triangular or keyhole-shaped.

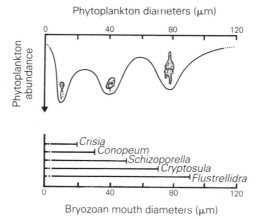

Figure 6.3 Mouth diameters of five bryozoans and hypothetical phytoplankton size distribution within the range of mouth diameters. (From Winston 1977b, courtesy of Academic Press.)

6.2.2 Estimating the size of lophophores and mouths

The number and length of tentacles and the diameter of lophophores can be approximated for most fossil bryozoans because they are related to sizes of zooecia and orifices. Winston (1981) obtained correlation coefficients between orifice width and lophophore diameter of 0.83 for cyclostomes and 0.78 for cheilostomes. Using the regression of lophophore diameter on orifice width for living bryozoans, as shown in Figure 6.4a, the lophophore diameter of fossil cyclostomes and cheilostomes can be predicted (Table 6.2, see also Winston 1981). Problems with this method of estimating lophophore diameter include widening of orifices in fossils by abrasion or solution, and non-preservation of orifices in cheilostomes that have partially membranous frontal walls.

Lophophore diameter in cheilostomes may also be predicted from zooecial length, zooecial width, and nearest-neighbor spacing of orifices (Fig. 6.4b–d). Correlation coefficients for the few values available are 0.66, 0.74 and 0.84, respectively. Apparently, orifice width and nearest-neighbor spacing are better predictors of lophophore size than are zooecial dimensions. Tentacle number generally increases with lophophore size and so also relates to size of skeletal structures (Fig. 6.5a–d). For cheilostomes, tentacle number also correlates better with orifice width and nearest-neighbor spacing of orifices than with zooecial dimensions.

Examination of these relationships for cyclostomes is complicated by the tubular shape of their zooids, commonly steep orientation with respect to the colony surface, and frequent presence of extrazooidal skeleton. Soft parts, such as tentacle number, are therefore better compared with orifice or living chamber width and length rather than with overall zooecial dimensions. Correlations of tentacle number with minimum living chamber diameters for cyclostomes are given in Figure 6.5e, f.

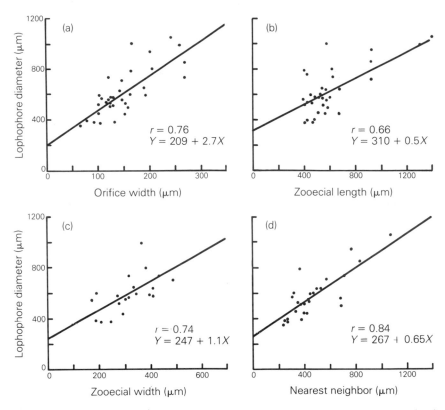

Figure 6.4 Relationship of lophophore diameters to selected skeletal measures in the Cheilostomata: (a) orifice width, (b) zooecial length, (c) zooecial width, and (d) nearest neighbor. Lophophore diameters are from Ryland 1975 and Winston 1979; skeletal measures are from specimens provided by D. Gordon and by J. D. & D. F. Soule.

Table 6.2 Estimates of mouth diameter and lophophore diameter of some fossil bryozoans (from Winston 1981).

	Mean orifice diameter (μm)	Estimated mouth diameter (μm)	Estimated lophophore diameter (μm)
Miocene stenolaemates (Vavra 1974)			
Bicrisina compressa	70	17	245
Exidmonea atlantica	110	25	384
Pleuronea pertusa	90	21	315
Eocene cheilostomes (Cheetham 1966)			
Microporina magnipora	180	64	759
Cribrilaria parisiensis	92	21	296
Escharella selseyensis	152	50	612

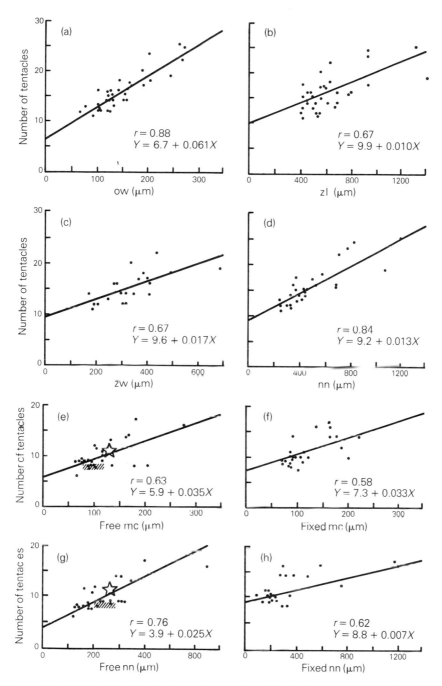

Figure 6.5 Relationship of tentacle number to selected skeletal measures in the Cheilosto-mata – (a) orifice width, (b) zooecial length, (c) zooecial width, (d) nearest neighbor – and Stenolaemata – (e) free-walled minimum chamber diameter, (f) fixed-walled minimum chamber diameter, (g) free-walled nearest neighbor, (h) fixed-walled nearest neighbor. Stars in (e) and (g) represent the trepostome *Tetratoechus crassimuralis*; ruled areas in (e) and (g) represent typical Paleozoic fenestrates. (Sources for (a)–(d) are the same as for Fig. 6.4; (e)–(h) from McKinney & Boardman 1985, courtesy of Olsen & Olsen.)

Fixed-walled cyclostomes (with the outer cuticle fixed against part or all of the skeletal wall) commonly have widely separated, apparently independently feeding polypides. Free-walled cyclostomes (with the outer cuticle free from the skeletal wall except where the colony is in contact with the substratum) typically have more closely spaced, more regularly patterned distributions of orifices than are found in fixed-walled forms. Because polypides in many fixed-walled species apparently function independently of each other, one would expect that features of the lophophores, including tentacle number, would correlate relatively poorly with nearest-neighbor spacing of orifices. In contrast, the closely spaced, patterned distribution of orifices in free-walled species would suggest relatively good correlation of lophophore features with nearest-neighbor spacing of orifices. This seems to be the case. For free-walled cyclostomes, $r = 0.76$ ($P < 0.01, N = 22$), and for fixed-walled cyclostomes, $r = 0.62$ ($P < 0.01, N = 21$) (Fig. 6.5g, h). There is even better correlation ($r = 0.85, P < 0.01, N = 11$) between tentacle number and nearest-neighbor spacing in cyclostome species that have compact clusters of free-walled zooids with fixed-walled spaces separating the clusters (McKinney & Boardman 1985).

One specimen of a Paleozoic trepostome with unusually well-preserved organic remnants shows similarities in numerous zooids to various soft parts in modern cyclostomes (Boardman & McKinney 1976). In one zooid there are 11 organic remnants interpreted as tentacles (Fig. 6.6), which places this

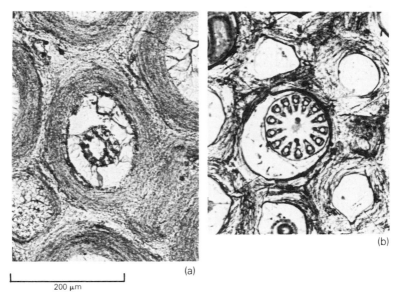

(a)

(b)

200 μm

Figure 6.6 (a) Cross sectional cut of red-brown granules, apparently representing tentacles and tentacle sheath, centered in a zooecium of the Paleozoic trepostome *Tetratoechus crassimuralis*; and (b) cross section of tentacles and tentacle sheath in a living heteroporid cyclostome for comparison. (From Boardman & McKinney 1976, courtesy of the Society of Economic Paleontologists and Mineralogists.)

Paleozoic species almost directly on the lines of regression of tentacle number on living chamber minimum diameter and nearest-neighbor spacing for extant free-walled cyclostomes (Fig. 6.5e, g).

Diameters of the distal tubes of Paleozoic fenestrates are typically about 80–120 μm, which is approximately the diameter of living chambers of extant cyclostomes with eight tentacles (Fig. 6.5e, f). In addition, nearest-neighbor spacing in Paleozoic fenestrates is typically between 200 and 300 μm, which falls within the range of nearest-neighbor spacing for living cyclostomes (Fig. 6.5g, h). Diverse fenestrates have "stellate" structures (Figs. 6.7, 6.19) at the apertures, consisting of vague to pronounced scallops around the margin, with shallow septa between adjacent scallops pointed toward the middle of the aperture. In almost every case there are eight scallops and intervening septa. The scallops are most reasonably interpreted as positions for exsertion of tentacles, suggesting that eight was the number of tentacles in lophophores of these Paleozoic fenestrates.

Mouth size in fossil bryozoans may be estimated from orifice size, zooecial length, and, in multiserial forms, nearest-neighbor spacing of apertures. Winston (1981) demonstrated correlation coefficients between orifice width and mouth diameter to be 0.96 for living cyclostomes and 0.78 for living cheilostomes. She then used regression of mouth size on orifice

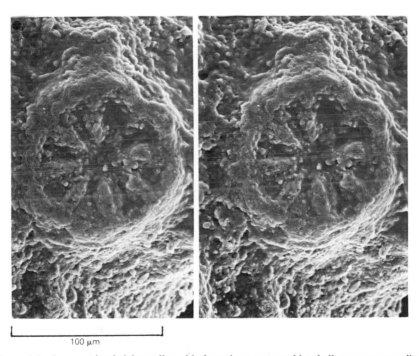

100 μm

Figure 6.7 Stereo pair of eight scalloped indentations separated by shallow septa extending axially from the perimeter of an aperture of the Carboniferous fenestrate bryozoan *"Penniretepora stellipora spinosa"* (Glasgow Art Gallery and Museum specimen 01–53vr).

size for living bryozoans to estimate mouth size in fossil bryozoans (Table 6.2). The same problems apply to estimating mouth size from orifice size as were listed for estimating lophophore diameter. However, nearest-neighbor spacing of orifices is also closely correlated with mouth size in cyclostomes ($r = 0.92$, $P < 0.01$, $N = 12$) and is not as readily affected by diagenesis or erosion as is orifice size.

6.3 Feeding currents and behavior

Where zooidal apertures are closely spaced, their polypides may cooperate to produce currents so that food can be effectively cleared from a larger volume of water than if zooids functioned independently. Fortunately for paleontology, the existence of such cooperative behavior is generally well reflected in local skeletal structures or in limitations on zooidal or colony size and form.

Colonies in which zooids feed cooperatively have the problem of development of excurrent regions while maintaining maximum possible area of the feeding surface. All solutions to this problem involve one or two of the following basic methods:

(a) flow of water to the edge of the continuous colony surface,
(b) development of nonfeeding, excurrent areas within the continuous colony surface area where filtered water moves away from the front of the colony,
(c) presence of perforations that permit filtered water to pass through the colony after it has passed through lophophores, and
(d) presence of channels on the colony surface, between elevated rows of apertures, where the water flows laterally to one or more areas of confluence and excurrent flow.

Each of these situations is considered below.

6.3.1 Feeding in Membranipora

Feeding currents and behaviors of species of *Membranipora* have been more extensively investigated than for any other group. Thin encrusting sheets of *Membranipora* on algae or hard substrata consist of rectangular to elongate hexagonal zooids that are contiguous and are packed in a rhombic pattern. Water is drawn toward the colony surface by filtering lophophores that are highly elevated. In colonies of only a few zooids, all the filtered water can flow in the space between the skeletal surface and the elevated lophophores and out from the colony margin. This is not the case, however, for large colonies, because the length of the margin of the colony grows proportionally with the radius, whereas the volume of water filtered increases with the square of the radius if most zooids feed simultaneously. Increasingly difficult mechanical problems caused by forcing all filtered water to the colony edge require that larger colonies have other means for expulsion of filtered water.

In *Membraniopa* colonies that consist of more than a few zooids, there are regularly spaced gaps (**chimneys**) in the distribution of protruded lophophores, through which filtered water escapes (Fig. 6.8, Banta *et al.* 1974). Most zooids in the colonies have equitentacled lophophores, but the chimneys are surrounded by zooids with obliquely truncate lophophores that have their longest tentacles bordering the chimneys. These lophophores also have longer introverts than are found in other zooids, and they tend to lean away from the chimneys, thereby enhancing excurrent flow. The chimneys are about 0.8 mm in diameter and are spaced about 2.3 mm from center to center, so that they occupy about 15% of the colony surface.

In a detailed study of flow over *Membranipora villosa* (possibly *M. membranacea*; see Harvell 1984a), Lidgard (1981) found that this species, which encrusts flat algal fronds in rapidly flowing water, has regularly spaced chimneys centered over degenerated zooids. Lophophore shapes and sizes are distributed as in other species of the genus, and the colonies feed from quiet water within the **boundary layer** (zone of reduced flow velocity along a stationary surface). Simultaneously, they eject filtered water through chimneys at sufficient force to propel it into the laterally flowing water above the boundary layer, thereby avoiding refiltering (Fig. 6.8).

Membranipora does not have pronounced skeletal reflections of chimneys. This condition is permitted by long, flexible introverts which allow tilting and lateral displacement of lophophores. However, there is apparently a correlation in *M. membranacea* and several similar encrusting species between chimney location and position of large zooids that precede zooid row bifurcation (Cook 1977a, Cook & Chimonides 1980). This pattern needs more rigorous verification, for row bifurcations are not evenly distributed over most colonies.

6.3.2 Types of feeding behavior

Winston (1978, 1979, 1981) recognizes the following categories of general behaviors and feeding currents:

(a) behavior individualized;
(b) polypides in fixed clusters, separated by colony skeleton;
(c) polypides grouped into temporary clusters, not reflected in skeletal morphology;
(d) polypides forming fixed clusters, not reflected in skeletal morphology;
(e) polypides in fixed clusters that are reflected in skeletal morphology, skeletal patterning either regular or irregular.

Morphological and physiological integration are correlated with behavioral integration as expressed in feeding currents among living cheilostomes (Winston 1978, Cook 1979, McKinney 1984, Coates & Jackson 1985).

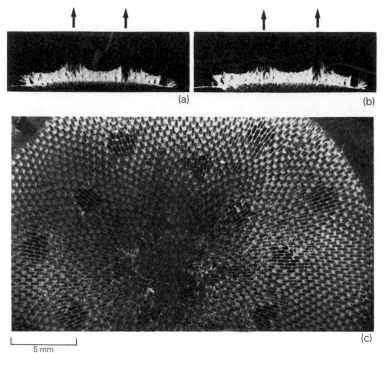

Figure 6.8 Chimneys in feeding *Membranipora villosa*; (a) lateral view showing introduction of dye above colony; (b) lateral view of dye being expelled from chimneys; (c) frontal view showing regularly distributed chimneys (dark areas). (From Lidgard 1981, courtesy of Olsen & Olsen.)

If Winston's feeding current types are arranged from least organized to most organized (i.e. types a, c, d, e and b, respectively), there is a corresponding increase in calculated values for states of colonial integration (Table 6.3, § 1.2, 4.2). Colonial integration for species in each of the better organized categories of feeding currents (types d, e, and b) are significantly higher than for categories with the more poorly organized feeding currents (types a and c) (One-way ANOVA least squares difference, $P<0.05$, $N=87$), but are not significantly different from one another.

The five types of feeding behaviors are considered further in the remainder of this section.

(a) Individualized behavior

Colony-wide currents are weak or absent in bryozoans with widely separated zooidal apertures. In such species there is sufficient space between expanded lophophores so that flow generated by each individual neither effectively reinforces nor interferes with that generated by its nearest neighbors. Lophophores in such bryozoans are equitentacled. Gymnolaemates of this type exhibit extensive scanning behavior, in which the

Table 6.3 Degree of physiological and structural integration in gymnolaemates with different colony current patterns. Data are for species listed by Winston (1978, 1979); values for integration are computed by assigning values of 0 to 1 for different states in six integration series (zooidal wall integrity, interzooidal connection of soft tissues, extrazooidal parts, astogenetic gradient, type of polymorphism, distribution of polymorphs).

Colony current patterns in order of organization	Colony integration			One way ANOVA LSD (0.05)				
	$\bar{X}$	range	N of species	A	B	C	D	E
A Independent zooidal currents	0.28	0.12–0.44	11					
C Currents by temporarily cooperating zooids	0.31	0.12–0.58	22					
D Currents by fixed groups of zooids, not reflected in skeleton	0.42	0.21–0.64	14	X		X		
E Currents by fixed groups of zooids, reflected in skeletons	0.46	0.25–0.58	9	X		X		
B Undirectional colony currents passing branches	0.50	0.35–0.64	19	X		X		

lophophore is rotated and redirected on its long introvert until a suitable orientation is attained. Cyclostomes, which have short introverts, do not show such behavior.

Lack of colonial currents has been noted only in uniserial and loosely multiserial colonies, such as encrusting uniserial (*Bowerbunkia, Nolella, Beania*), boring uniserial (*Terebripora*), erect uniserial (*Aetea*), and erect stoloniferous (*Amathia, Bowerbankia*) forms. Zooids in comparable fossil species (*Stomatopora, Corynotrypa, Voigtopora, Pyriporopsis*) probably also functioned individually, without producing colonial feeding currents.

(b) Polypides in fixed clusters that are separated by colony skeleton

These develop highly organized currents that flow through slots between elongate groups of zooids. Zooids are organized into narrow unilaminate branches (*Bugula, Phidolopora*), and probably similar currents develop in colonies with highly elevated thin, branch-like ridges in which zooids all face the ridge crest (*Telopora*). Nearest polypides from both branches beside each slot protrude into the overlying space and cooperatively draw water toward the colony surface (Fig. 6.9); filtered water then continues through the space between branches, moving away from the feeding surface.

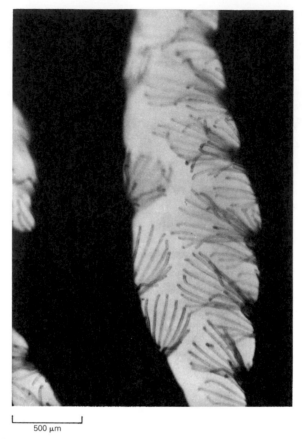

500 μm

Figure 6.9 Obliquely truncate lophophores of the cheilostome *Reteporellina evelinae* protruded over, and driving a strong current through, the narrow space between branches. (From Winston 1978, courtesy of the University of Miami, Rosenstiel School of Marine Science.)

(c) Polypides in temporary clusters

Some colonies with broad surfaces do not form chimneys, but rather have temporary clusters of a few zooids with circular lophophores whose location and orientation are determined by the first member of the cluster to emerge and to feed (Winston 1978, 1979). There is no skeletal indication of where the temporary clusters are likely to occur. These bryozoans generally have fewer zooids feeding at any given time than do those with better integrated feeding behavior.

(d) and (e) Polypides in fixed clusters

Excurrent chimneys develop on broad surfaces, such as on large branches or encrusting sheets, where most zooids feed simultaneously. As in *Membranipora*, most chimneys are bounded by zooids that lean away from the excurrent areas and that have obliquely truncate lophophores with longer tentacles adjacent to the chimneys. Regions between chimneys are uniformly covered by circular lophophores that are expanded to barely overlap and thereby fill the area. Chimneys are generally, but not always, circular, and are fairly uniform in size and spacing within a species (Banta *et al.* 1974, Cook 1977a). Chimney size and lophophore diameter are directly related in the small number of species examined (Winston 1979).

The long introvert in cheilostomes allows zooids to tilt their lophophores away from adjacent chimneys, and at least some taxa, such as *Membranipora*, have zooids with longer introverts bordering chimneys. It is therefore not necessary that chimney positions be recorded in skeletons of cheilostomes, although some (e.g., *Celleporaria, Hippoporidra*) have elevations regularly spaced at 2–3 mm intervals that coincide with chimneys (Winston 1978). In the absence of skeletal indications, it is apparently impossible to determine if a fossil cheilostome with a broad surface of contiguous zooids fed by means of temporary clusters of zooids or generated regularly patterned currents.

Chimneys in stenolaemates should be more likely reflected in skeletons than those in cheilostomes because the feeding position of lophophores is closer to the skeletal surface in stenolaemates. In stenolaemates, water passing through lophophores therefore would immediately come against the colony surface, which must, in turn, reflect paths of flow in its shape.

Fossil stenolaemates, and some fossil cheilostomes with broad surfaces, typically have regularly spaced raised or lowered areas, termed **maculae**. The maculae are of the order of a fraction of a millimeter to about 3 mm in diameter and most are spaced at intervals slightly greater than twice their diameters, so that they occupy from less than 10% to over 20% of the surface area (Fig. 6.10). The maculae apparently mark the positions of excurrent chimneys (Banta *et al.* 1974, Taylor 1975, 1979c, Boardman & McKinney 1976), as their approximate size, spacing, and per cent surface area that they occupy resemble those of chimneys observed in living cheilostomes. Their mode of construction is highly variable, including masses of extrazooidal skeleton, large or small heterozooids, and normal-sized zooecia that form elevations or depressions relative to the level of the general colony surface. Such diverse origins but regular surface disruption suggest that the primary function of maculae is related to a surface phenomenon: separation of distinct incurrent and excurrent areas.

Two morphologies of maculae directly reflect centripetal flow. The Ordovician cystoporate *Constellaria* and related forms have star-shaped maculae that consist of depressed central areas of interzooidal skeleton that extend outward as channels between radially disposed double rows of zooecia (Figs. 6.10d, 6.11a). Individual maculae in *Constellaria* resemble excurrent channel systems in small mound-shaped cyclostomes and have the same radially bifurcating pattern as exhalent channels in sclerosponges, although converging channels in sclerosponges are internal (see Hartman

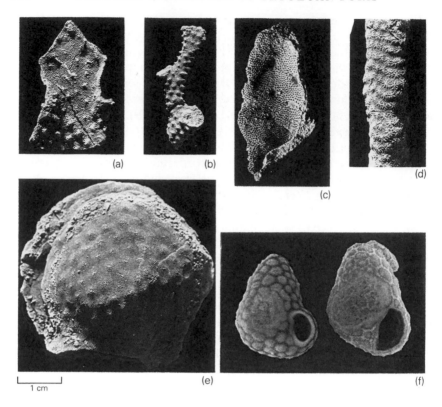

1 cm

Figure 6.10 Maculae in Paleozoic stenolaemates, including the bifoliate cryptostome *Arthrophragma* (a); the trepostomes *Parvohallopora* (b) and *Tabulipora* (e); the cystoporates *Fistulipora* (c) and *Constellaria* (d); and in the Recent cheilostome *Hippoporidra* (f), a commensal with pagurid crabs in gastropod shells, having two different cnidarian-mimicking color morphs with patterns centered on the maculae.

& Goreau 1970). In some cyclostomes and cystoporates, zooecia immediately around maculae are angled away from the macular center (Figs. 6.10c, 6.11b), which suggests that the lophophores when protruded were positioned to enhance flow toward maculae.

Most maculae are roughly circular, but there are intergrading cases to distinct ridges (for elevated maculae) or channels (for depressed maculae). Colonial current patterns can be estimated for these other types of surface patterns by centering excurrents on the ridges or channels and by reconstructing incurrent water over intervening autozooecial areas.

Although maculae probably functioned primarily as excurrent areas, other functions are not precluded. They are the sites of male polymorphs in *Hippoporidra* (Banta *et al.* 1974, Cook 1977a), which is a suitable situation for dispersal of sperm, and some maculae are also sites of increase in number of zooids (Anstey & Delmet 1972, Anstey *et al.* 1976).

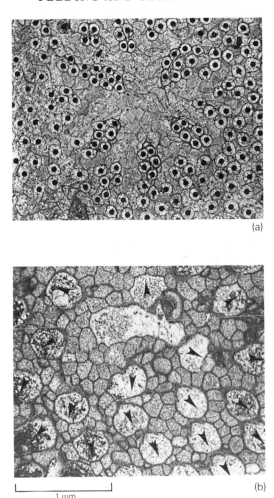

(a)

(b)

1 mm

Figure 6.11 Near-surface tangential peels of cystoporates that have autozooids (marked by black spots and arrows) organized differently around maculae: (a) biserial clusters of autozooids radiating from center of macula in *Constellaria*; (b) *Fistulipora*, which has large autozooids with thickened, crescent-shaped portions of walls (lunaria) adjacent to the central macula, thereby making a roughly radial pattern.

6.4   Feeding and colony form

Some highly distinctive modes of feeding are closely tied to particular colony forms. Similar patterns of feeding and morphology have evolved independently many times within various stenolaemate orders and within the cheilostomes.

6.4.1 Mound-shaped colonies

Single chimneys exist in the center of small mound-shaped stenolaemate colonies with diameters up to 1 cm, such as *Disporella* and *Plagioecia* (Figs. 6.2, 6.12a). In such colonies water is drawn in across the colony surface and converges toward the colony center. Such convergent or centripetal flow is confined to channels between radiating rows of zooids, because there is no room for flowing water between lophophore bases and apertures of the zooids. Filtered water flows along the channels toward the central colony open area or summit, where it is expelled.

Colonies (Fig. 6.12a) are commonly zoned as follows:

(a) a peripheral growing edge characterized by an ontogenetic gradient in polypide and lophophore size;

(b) a zone where zooids have elongate peristomes so that lophophores are protruded above the region of previously filtered, centripetally-flowing water; and

(c) a central zone where peristomes have broken off, orifices are closed by terminal diaphragms, and active polypides are lacking (Silén & Harmelin 1974).

The entire central zone serves as a chimney, with velocity of excurrent flow inversely related to proportion of area in the central region relative to area of the peripheral zone of actively feeding polypides.

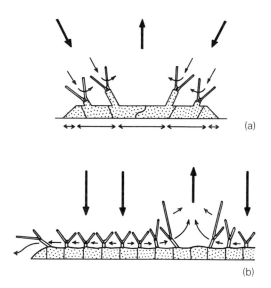

Figure 6.12 Patterns of water flow in (a) small, ontogenetically zoned, mound-shaped cyclostome colonies and (b) sheet-like encrusting cheilostome colonies. (Modified from Taylor 1979c and Cook 1977a, courtesy of the International Palaeontological Association and *Cahiers de Biologie Marine*.)

6.4.2   Continuous sheets and lamellar colonies

Colonies that develop as thin encrusting sheets may initially, when they consist of only a few tens of zooids, expel all filtered water along the colony periphery, as in small colonies of various species including *Flustrellidra hispida* and *Cleidochasma contractum* growing on bivalve shells (Cook 1977a). When colonies are larger, filtered water in the middle portions of the sheet is exhausted through chimneys, while that filtered near the colony margin is forced out along the periphery (Fig. 6.12b).

"*Hippodiplosia*" *insculpta* expands as a sheet along narrow cylindrical substrata, such as algae and gorgonians, until the growing edges of one or more colonies meet, at which point the sheets grow back-to-back, away from the original substratum. These bilamellar portions comprise most of the colony area. They are covered by expanded equitentacled lophophores, making a continuous inhalent area, with filtered water expelled along the colony edge (Fig. 6.13). Lophophores along the colony edge are obliquely truncate, with their longer tentacles extending beyond the colony margin. The marginal expulsion of filtered water restricts width of "*Hippodiplosia*" colonies to a few millimeters because, inasmuch as no chimneys are present, lophophores of the innermost zooids must work against an increasing resistance as the colonies widen, "and this can be possible only to a certain

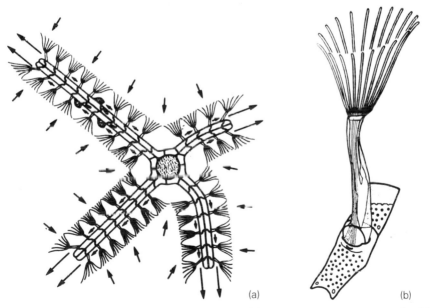

(a)   (b)

Figure 6.13 Colony-wide water flow in "*Hippodiplosia*" *insculpta* as seen in cross section (a) consists of areas of inflow across the flat surfaces, lateral flow at the level of introverts, and outflow along the keel-shaped colony edges, where (b) polypides have extra length and obliquely truncate lophophores. (From Nielsen 1981, courtesy of Marine Biological Laboratory, Denmark.)

limit" (Nielsen 1981, p. 97). Were chimneys present, size of colonies presumably would not be so restricted.

6.4.3   Adeoniform colonies

Pattern of flow in adeoniform colonies (Figs. 3.4, 6.14), and also in arborescent bifoliate colonies that are articulated, has not been reported, but should be like that described above for *"H." insculpta* if all or most zooids feed simultaneously. This would limit branch width for each species, unless chimneys were present. Indeed, living and fossil adeoniform colonies are characterized by rather uniform branch width (Fig. 6.15a, McKinney 1986b). Although branch thickness in many such species increases with branch age, branch *width* remains largely unchanged (Cheetham *et al.* 1981).

Adeoniform growth has been common since the beginning of the fossil record of bryozoans (Fig. 6.16), and has evolved independently within the orders Cyclostomata, Cystoporata, Trepostomata, Cryptostomata, and Cheilostomata, often several times within the same order (e.g., *Membranipora savartii*, *Metrarabdotos tenue*, and *Adeonella platelea*, within the

5 mm

Figure 6.14   Adeoniform colony of the Ordovician cryptostome *Graptodictya perelegans*. (Photograph courtesy of O. L. Karklins.)

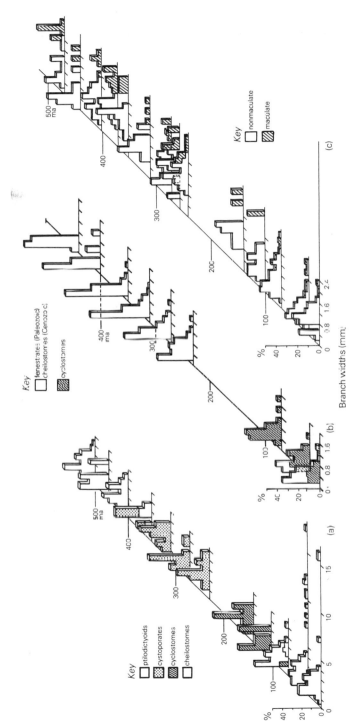

Figure 6.15 Histograms of (a) adeoniform, (b) nonmaculate radial, and (c) unilaminate branch widths from Ordovician to Recent. Data are from 63 taxonomic monographs. (From McKinney 1986b, courtesy of the Royal Society, London.)

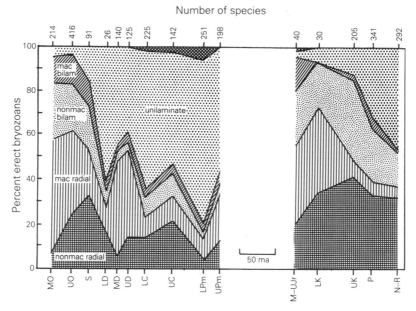

Figure 6.16 Plot of relative abundance of growth forms among erect bryozoans, from the time of first abundance to Recent. The unlabelled growth form with low abundance from Upper Devonian to Lower Cretaceous has narrow bilaminate ("eccentric bilaminate" of McKinney 1986b) branches with pyriform cross sections and zooidal apertures deflected toward the narrow edge. Key to abbreviations: mac = maculate, nonmac = nonmaculate, bilam = bilaminate. Abbreviations along lower edge are for Phanerozoic systems and series. Data are from 63 faunal monographs. (From McKinney 1986a, courtesy of The University of Chicago Press.)

Cheilostomata). In all these groups, maximum branch width has generally not exceeded 3–4 mm (Fig. 6.15a). Where maximum branch width is greater than 5 mm, maculae are present (Fig. 6.17) except in some cheilostomes, where chimneys are not necessarily reflected in skeletons. The presence of maculae on broad bifoliate branches and sheets, and their absence on adeoniform branches, strongly supports Nielsen's suggestion that there is a maximum possible distance from colony center to excurrent regions along the edge of bifoliate colonies where chimneys are lacking. There also seems to be a minimum diameter for development of maculae on cylindrical branches (Fig. 6.15b).

6.4.4   Unilaminate arborescent colonies

Bryozoans constructed of narrow unilaminate branches have zooids oriented towards only one side. When their zooids feed, their obliquely truncate lophophores generate a current that is drawn towards and then passes by the branches (Fig. 6.18). Existence of such obliquely truncate

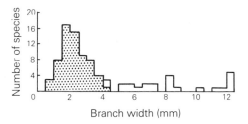

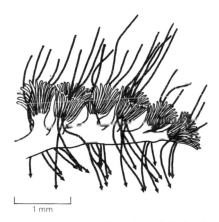

Figure 6.17 Branch widths of Ordovician bifoliates. Patterned portion represents branches that lack maculae, including adeoniform and articulate colonies with ribbon-shaped branches, and open portions represent colonies that bear maculae, including broad adeoniform and eschariform. $N = 89$. (Measured from illustrations in Ulrich 1893, Nekhoroshev 1961, Astrova 1965, Karklins 1969.)

Figure 6.18 Drawing of the path of silt particles through lophophores and past a branch in a colony of *Bugula neritina*. (From McKinney *et al.* 1986a, courtesy of the International Palaeontological Association.)

lophophores and through-flowing currents was hypothesized by Cowen & Rider (1972) and subsequently confirmed by Cook (1977a), McKinney (1977b), and Winston (1978, 1979).

Lophophores of zooids along branch margins have their longest tentacles extending laterally away from the axial planes of their branches. Where the branches are closely spaced, lophophores from zooids of adjacent branches fill the open space between, forming filtration sheets that pass the filtered water through spaces between branches. In many cases zooids are positioned so that they fill the space with a continuous sheet of lophophores. For example, at each branch-like origin in *Bugula turrita*, one zooid is 50% longer than usual so that orifices in the two zooid rows along the branch alternate from side-to-side, producing a rhombic spacing of lophophores. Zooecial apertures in many Paleozoic fenestrates are similarly offset, so that apertures in several adjacent branches form a rhombic pattern (Fig. 6.19).

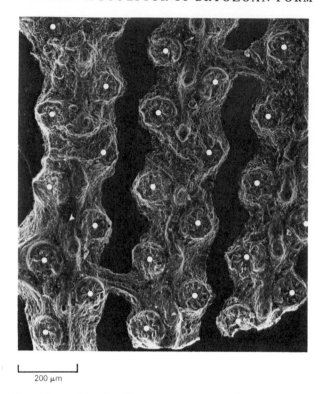

200 μm

Figure 6.19 Rhombic spacing of stellate apertures across three biserial branches of the Carboniferous fenestrate *"Actinostoma fenestratum"*.

Unilaminate branches are generally narrow (Fig. 6.15c). Most species have only two rows of zooids, but up to 12 rows are known in *Retiflustra cornea* (Harmer 1926), and 3–8 rows are common in Paleozoic fenestrates. It appears that a trade-off exists between not having to divert any colony surface area to excurrent chimneys and maximum attainable branch width.

Taxa with unilaminate, regularly spaced branches that form a continuous surface of expanded lophophores have evolved independently in the fenestrates (the entire order), cyclostomes (*Idmidronea, Hornera*), and cheilostomes, in which they are represented by taxa with rigid skeletons (*Sertella, Phidolopora*) and by others with flexible skeletons (*Bugula*). Compared with other erect forms, unilaminate branched colonies seem to have the advantages of a relatively light skeleton and a continuous feeding surface with the possibility of reduced resistance to outflow of filtered water by streamlined and reduced branch cross sections. Their primary disadvantage is their vulnerability to breakage. Their advantages must outweigh their disadvantages, for they have tended to increase dramatically in diversity relative to other erect growth forms in both the Paleozoic and post-Paleozoic (Fig. 6.16, McKinney 1986a).

Feeding from self-induced currents that flow through filtration sheets requires considerable coordination of development of separate portions of colonies. Sets of branches side-by-side must grow at comparable rates, and none may bend out of the plane of neighboring branches by more than a certain amount if the filtration surface is to function as a single unit. In addition, close impingement of any two sheets of zooids would affect efficiency and eventually any ability to function for both sheets involved. Similarly, a filtration sheet cannot generate through-flowing currents if it is blocked by a substratum, excessive calcification, or any other obstruction that would block spaces between branches.

Bugula turrita grows as spiralled filtration sheets with superposed whorls spaced 2–3 mm apart. There are two rows of zooids per branch, and frontal surfaces of branches face the distal end of the spiral (Fig. 3.12a). Unidirectional currents pass downward through the colonies (Cook 1977a, Winston 1979, McKinney *et al.* 1986a). Water passes rapidly through lophophores and between branches, then drifts more slowly past the reverse side of the branches before being processed through the next whorl down. Only the distal branches and outermost few zooids of more proximal branches contain active polypides. Apparently, the metabolic cost of maintaining active polypides in the colony's interior is greater than the gain of food from repeatedly refiltered water to which they would have access, while zooids at ends of branches have access to a combination of filtered and "new" water.

Archimedes (Fig. 3.12) is a Paleozoic genus of fenestrates whose growth and distribution of branches is very similar to *Bugula turrita* (Ch. 3, McKinney 1980). An additional similarity is that most inner and proximal zooids were nonfeeding, each being capped by an imperforate terminal diaphragm. *Archimedes* is therefore inferred to have generated currents flowing through the colony as in living *B. turrita*.

The lyre-shaped fenestrates (Fig. 6.20) appear to have used the same basic plan of closely spaced unilaminate branches to feed within the boundary layer in regions of pronounced unidirectional ambient current (McKinney 1977b). Following an initial period of upright growth as a planar fan with edges progressively bowed downcurrent, as in *Gorgonia* and other planar octocorals (Wainwright & Dillon 1969, Leversee 1972, Rees 1972, Velimirov 1973, Muzik & Wainwright 1977), most colonies toppled in the downcurrent direction. They rested on the edges of the bowed fan and continued to grow down current, with the upcurrent end shaped like a parabola and progressively more heavily calcified. Colonies fed from water that flowed over the exposed surface; passive flow was generated by vacuum on the downcurrent end of the colony, and ciliary pumping reinforced the flow of water past lophophores and branches, into the space between colony and substratum from which it was vacuumed downstream through the open end. The cheilostome *Canda*, which grows in quiet water on the undersides of platy corals, has a growth form somewhat similar to that of Paleozoic lyre-shaped fenestrates. Most colonies are fan-shaped, bent over parallel with the substratum; zooids are arranged in two rows on branches and face outwardly. Their expanded lophophores generate a current that flows between branches, toward the reverse side of the fan (Winston 1981).

1 cm

Figure 6.20 Colony of *Lyropora* sp. (Carboniferous, Viséan–Namurian, Illinois), with thick extrazooidal skeleton along proximal margin at left; distal (growth) direction is toward right.

7 Encrusting growth: the importance of biological interactions

Community composition on large, stable substrata such as pier pilings, reefs, and rock walls is generally controlled by interactions among established organisms. In contrast, community composition on small or ephemeral substrata such as algae, dead shells, and cobbles is determined more by rates of recruitment and habitat selection by larvae than by post-recruitment interactions. Bryozoans are generally poor competitors for space and lose most encounters with other clonal encrusting organisms such as sponges, cnidarians, and ascidians (Gordon 1972, Kay & Keough 1981, Jackson & Winston 1982, Russ 1982, Keough 1984a, b). Most bryozoans are also comparatively slow to regenerate injuries due to physical disturbances or predators (Jackson & Palumbi 1979, Ayling 1981, Palumbi & Jackson 1982, Ayling 1983). Consequently, encrusting bryozoans tend to occur more commonly and abundantly on ephemeral than on more permanent substrata (Fig. 7.1).

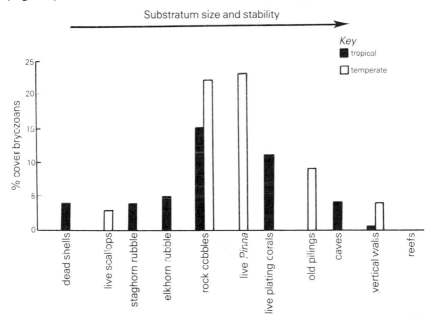

Figure 7.1 Per cent cover of bryozoans on substrata of different size and physical stability. (Data from: Vance 1979, Ayling 1981, Liddell *et al*. 1984, Jackson unpublished, and references 7, 8, 10, 12, 14 for Table 7.2.)

Encrusting bryozoans on ephemeral substrata are the weedy species of Chapter 5, which comprise the great majority of all Recent bryozoan species. There are, however, striking exceptions to the general exclusion of encrusting bryozoans from large stable substrata. These exceptional species, which are exclusively cheilostomes, have in common a greater flexibility of modes of growth (budding) than most other bryozoans, which increases their chances of survival in interactions with other sessile organisms and predators (Lidgard & Jackson 1982, Lidgard 1985a, b). In environments where these species prevail, recruitment is slow and plays a relatively minor role in determining community composition compared with the importance of biological interactions.

In this chapter we review the ecology of encrusting communities on substrata of differing stability with emphasis on ecological processes most important in determining the types and abundances of bryozoans found, and the morphological innovations which have allowed some cheilostomes to attain great abundance in environments unfavourable to most bryozoans. Development of progressively more versatile and integrated modes of budding has resulted in a major evolutionary transition in the proportions of encrusting cheilostome species with different growth patterns from the Late Cretaceous to the present (Lidgard 1986). The dramatic decline in abundance and diversity of encrusting cyclostomes throughout the Tertiary may be another consequence of this trend.

7.1   Short-lived or unstable substrata

Fronds of many seaweeds last for a year or less, even those of plants which may survive for many years. The same is true for the blades of seagrasses. Thus from the perspective of an encrusting organism, these plant parts are ephemeral substrata for all but the shortest-lived species. On the other hand, plants like *Macrocystis*, *Fucus*, *Posidonia*, and *Thalassia* are extremely abundant and predictable substata, as evidenced by the numerous epiphytic species which have evolved highly specific relationships with their hosts (Hayward 1980, Seed & O'Connor 1981). Examples include habitat selection by larvae for different algal species (Ryland 1974) and for different positions (and therefore ages) on their fronds (Stebbing 1972, Hayward & Harvey 1974) and directional growth of colonies towards younger parts of fronds (Ryland & Stebbing 1971). In contrast, other bryozoans that occur on the same plants but which lack such specific behaviors tend to occur on a much wider variety of substrata (Ryland 1962).

More is known about factors controlling the distribution and abundance of *Membranipora membranacea* and other bryozoans on fronds of the giant kelp *Macrocystis pyrifera* off southern California than for any other such association (Fig. 5.7, Bernstein & Jung 1979, Yoshioka 1982a). The most striking aspect of this canopy community is the rapid variation in abundance of characteristic species, which may change more than 1,000-fold in a few months. *Membranipora* has long-lived planktonic larvae which recruit throughout the year. Regression analysis shows that variations in larval abundance (known from plankton tows) and seawater temperature (which strongly affects larval behavior) can explain 79% of the variation in recruitment onto blades. This recruitment, and predation on colonies by fishes and nudibranchs, are the primary factors controlling population sizes and fluctuations of *Membranipora*. However, predators tend to be attracted only by dense populations of *Membranipora*, when competition for space is also more likely to occur. These density-dependent factors probably act to maintain *Membranipora* populations within comparatively small limits from year to year, despite their potential for enormous short-term variations. The strength of the data on which this view is based is evident from Yoshioka's (1982a) ability to predict quite well long-term trends in population size (Fig. 7.2).

Few bryozoan epiphytes have been investigated thoroughly enough to assess so well the relative importance of different factors controlling their distributions. It is clear, however, that both the longevity and rigidity of their substrata play an important role. For example, *Macrocystis* blades are flexible and rarely survive for six months (North 1961), whereas those of the sympatric red alga *Rhodymenia californica* are comparatively rigid (Abbott & Hollenberg 1976), survive longer, and support a more dense and diverse epiphyte community (Bernstein & Jung 1979). Sponges and colonial ascidians are common on *Rhodymenia* and competition is intense, as evidenced by multiple layers of epiphytic overgrowths.

Dead shells of most molluscs and sea urchins provide abundant unstable (easily moved about or buried) and ephemeral (easily broken) substrata,

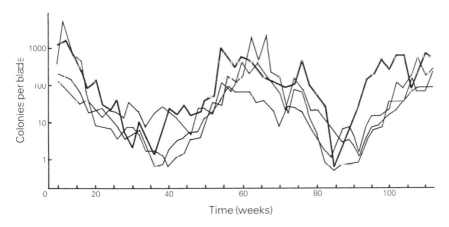

Figure 7.2 Observed (heavy line) and predicted (light lines) densities of colonies of *Membranipora membranacea* on kelp. (From Yoshioka 1982a, courtesy of the Ecological Society of America.)

commonly inhabited by a large proportion of bryozoan species (Ch. 4, Fig. 7.3, Eggleston 1972a, Harmelin 1977), but almost nothing is known about the dynamics of communities encrusting shells or the factors responsible. Bryozoans and solitary animals such as serpulids are the most abundant animals encrusting shells; sponges and ascidians are generally rare or absent. The encrusted area of shells is usually less than the space available, and encounters between organisms infrequent. Community composition almost certainly reflects rates of larval recruitment and destruction of shells more than effects of biological interactions.

Commonly, bryozoans may even encrust substrata as unstable as single sand grains, where they live interstitially within the top few centimeters of sandy shelf sediments (Figs. 7.4, 9.3, Håkansson & Winston 1985, Winston & Håkansson 1986). Bryozoans and Foraminifera are the most abundant encrusting organisms on sand, but cover of grains is at most a few per cent, due partly to abrasion by moving grains during rough weather. Length of life is unknown but seems very short as evidenced by the tiny size of colonies (usually less than 25–30 zooids), extensive partial mortality of colonies due to abrasion, and the presence of about 20 dead for every living colony. Larvae settle into concavities of grain surfaces where they are almost certainly better protected from abrasion. There is also a suggestion of

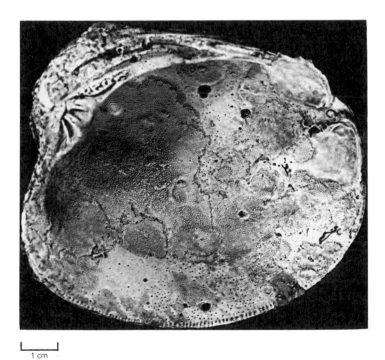

1 cm

Figure 7.3 Dead bivalve shell from near Beaufort, North Carolina, heavily encrusted by bryozoans.

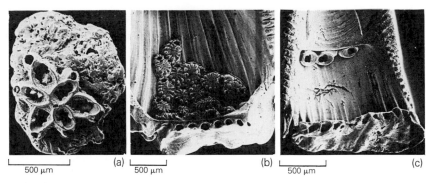

500 μm   (a) 500 μm   (b) 500 μm   (c)

Figure 7.4 Bryozoans on sand grains from Capron Shoals, Florida: (a) *Cupuladria doma*, (b) *Cribrilaria innominata*, (c) *Cymulopora uniserialis*. (Photographs courtesy of J. E. Winston.)

habitat selection for particular sizes of grains. Like many minute rooted species (§ 9.2), colonies are sexually precocious, forming ovicells after attaining only 3–6 zooids. Reproduction and recruitment are almost entirely sexual, although there is one example of a biserial membraniporid that bridges gaps between sand grains by growth of uncalcified nodes. These presumably break to form separate colonies by fission. Although evidence is circumstantial, it all points toward the community composition of sand grains being controlled by rates of larval recruitment and abrasion.

7.2 Substrata of intermediate longevity and stability

Pinna is a large, semi-infaunal bivalve mollusc that often lives for 5–10 years or more. Shells project well above the sediment where they typically support a diverse encrusting community (Fig. 7.5a, b). The epifauna of *Pinna bicolor* in South Australia is the best understood (Table 7.1, Kay & Keough 1981, Keough & Butler 1983, Keough 1984a, b). Exposed surfaces of adult shells available for settlement average 150–180 cm². Bryozoans are the most abundant epifauna; they average about 23% cover, and occur on almost all shells. Sponges average only 10% cover and occur on only 10% of the shells. Serpulids are ubiquitous but their cover is very low. Colonial ascidians are rare. Total cover by organisms averages less than 40%. Bare space is almost always present (Fig. 7.5a), except for the few shells overgrown by sponges or colonial ascidians (Fig. 7.5b). These groups, which are by far the best overgrowth competitors for space, exhibit very low levels of larval recruitment, and thus encounter few shells. Where they are absent bryozoans predominate. Bryozoans commonly overgrow serpulids, and occasionally encounter and overgrow each other (Fig. 7.5c). However, even bryozoan recruitment is low compared to that on kelps, averaging only 0.1–1 individual per shell in two months for most encrusting species. This low recruitment probably results from the comparative isolation of *Pinna* shells

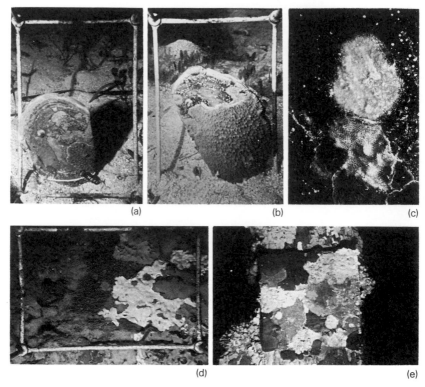

Figure 7.5 Encrusting communities at Edithburgh, South Australia. (a) *Pinna bicolor* encrusted by bryozoans *Schizoporella schistosoma*, *Celleporaria* sp., and a tag; (b) *Pinna* with *Schizoporella* colonies and spirorbids being overgrown by a large sponge; (c) close-up of the surface of a *Pinna* shell with spirorbids and three bryozoans in contact, on the left *Parasmittina raigii*, and two species of *Celleporaria*; (d) surface of piling cleared about 15 months previously (area delineated by nails within the frame). All of the organisms are sponges and colonial ascidians except for two small bryozoans, most likely *Membranipora perfragilis*, at the center of the clearing. (e) Isolated settlement panel (625 cm^2) placed on piling about 15 months previously. Most of the surface is overgrown by bryozoans whereas the piling is dominated by sponges and colonial ascidians. (Photographs courtesy of M. J. Keough.)

from large breeding populations of bryozoans, and the extremely limited dispersal capabilities of brooded bryozoan larvae (§ 5.5.1). In contrast, serpulids recruit heavily onto all shells. Caging experiments demonstrate that predation by fishes may exclude colonial ascidians when they occasionally recruit in noticeable numbers. However, those ascidians which do survive tend to senesce within 18 months anyway. Thus, community structure on *Pinna* shells seems largely controlled by rates of larval recruitment.

Similar assemblages occur on the undersides of rock cobbles (Osman 1977). Stability of these substrata varies with size and exposure to waves. Large rocks (>0.1 m^3) are probably stable for more than one year, even at

Table 7.1 Ecological characteristics of the principal groups of encrusting epifauna on shells of *Pinna bicolor* and on pier pilings at Edithburgh, South Australia (Kay & Keough 1981, Keough 1983, 1984a,b).

	Sponges	Ascidians	Bryozoans	Serpulids
Growth mode	clonal	clonal	clonal	aclonal
Maximum size	$1\,m^2$	$1\,m^2$	$50\,cm^2$	$0.1\,cm^2$
Cover (%)				
pilings	52–62	2–14	8–10	1
Pinna	10.0	0.1	23.0	1.2
Shells colonized (%)	10	6	97	>98
Mean % cover when present	44	78	23	1.2
Overgrowth ability	high	very high	intermediate	poor
Larval recruitment	rare	low, sporadic	moderate, consistent	high, consistent
Clonal recruitment	fast	fast	slow	none

moderately exposed sites, whereas those much smaller may be constantly overturned and abraded. In accordance with this decreased stability, small rocks are inhabited mostly by barnacles and serpulids which settle in enormous numbers, whereas larger rocks are dominated by bryozoans and sponges which recruit in much lower numbers. Subtidally, large rocks are apparently moved little and are commonly entirely overgrown by a single species. The commonest bryozoan to do this is *Schizoporella errata* which may form many circum-encrusting layers in which the remains of overgrown organisms are clearly seen (Fig. 7.6).

7.3 Long-lived, stable substrata

Boulders, pier pilings, rock walls, and many reef corals provide stable substrata that may persist largely unchanged for decades or centuries. These are characteristically overgrown by clonal algae, sponges, cnidarians, and ascidians, the proportions of each group depending on the relative intensity of predation (grazing), wave action, light, and the types and abundance of suspended food (Birkeland 1977, Adey 1978, Ayling 1981, Kay & Keough 1981, Jackson & Winston 1982). Encrusting bryozoans are often common in these environments, but rarely dominate them because of their generally poor abilities to compete for space and regenerate injuries.

Most of the pier pilings at Edithburgh, South Australia have been submerged for at least 68 years (Keough & Butler 1983), and are encrusted by a community that differs strikingly from that on *Pinna* shells nearby (Fig. 7.5d, e, Table 7.1). Organisms encrust about 75% of the surface of the pilings. Community composition is quite stable (>5 years' observations), even at the species level, because most of the dominant species are long-lived and can grow clonally. This stability exists in spite of considerable turnover of encrusting biota on the pilings (primarily due to overgrowth, but

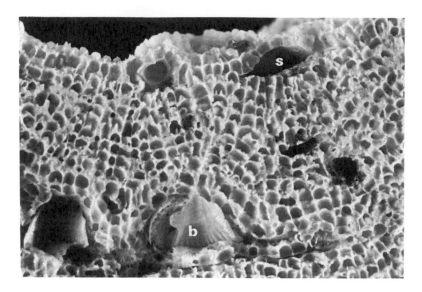

Figure 7.6 Cross section through frontally budded layers of *Schizoporella errata* at Woods Hole, Massachusetts, showing overgrown barnacles (b) and serpulid worm (s). Zooids are approximately 0.4 mm thick. (Photograph courtesy of Karl Kaufmann and Plenum Press.)

also to predation and senescence, mostly of colonial ascidians) which averages 10–20% every three months. When new space becomes available on a piling it is rapidly colonized by larvae and by encroachment of neighboring clonal organisms (Fig. 7.5d, Kay & Keough 1981, Keough 1984a). Larval recruitment is primarily by the relatively short-lived members of the community, including solitary animals (serpulids, barnacles) and many bryozoans (Fig. 7.5e). Bryozoan recruitment is as much as several hundred times higher than on *Pinna*, which correlates well with the much greater local abundance of parental stock under the pier (§ 5.5.1). Larval recruitment by sponges and colonial ascidians on the pilings is almost nonexistent, and has not been observed for many species. However, most of the new larval recruits are soon overgrown through clonal encroachment by these competitively dominant animals, whose growing edges may invade newly cleared patches as fast as 15 cm in 30 days. One striking evolutionary consequence of this pattern is habitat selection by bryozoan larvae for small substrata, which are much less likely to have been colonized by dominant competitors than are large substrata. This is evidenced by differential recruitment rates onto different sized panels attached to pilings.

Similar patterns are evident for pilings versus settlement panels (similar size to *Pinna* shells) at Beaufort, North Carolina (Sutherland & Karlson 1977, Karlson 1978, Sutherland 1978, 1981). A major difference at Beaufort, however, is extraordinarily high rates of larval recruitment, perhaps sustained by the high productivity of the estuary, which may overpower the defense mechanisms of established clonal animals. Another difference between the sites is the extreme seasonal range of water temperatures at Beaufort, which is coupled with senescence or dying back of many encrusting animals. Thus, turnover of space occurs for more reasons than at Edithburgh, and the community is more dynamic.

The undersurfaces of foliaceous corals on Caribbean reefs support a diverse encrusting community of algae, sponges, and bryozoans (Figs. 5.9, 5.17, Jackson et al. 1971, Jackson & Winston 1982, Palumbi & Jackson 1982, Jackson 1984, Jackson & Hughes 1985, Jackson & Kaufmann 1987). Overgrowths are extremely frequent at all depths, and predation by sea urchins and fishes is also common in depths less than about 15 m. Turnover of organisms under corals is fast, averaging about 50% per year. Nevertheless, community composition on the best-studied reef has not varied appreciably for at least ten years. The life histories of dominant organisms under corals are similar to those on pilings, but scaled down to the smaller dimensions of the under-coral habitat (usually 0.1–0.25 m^3). For example, most sponges under corals are only 1–3 mm thick (as opposed to several centimeters thick on pilings or open reef surfaces), and cannot so readily overgrow the surfaces of bryozoans which are of comparable thickness. This smaller spatial scale, plus growth of corals which provides new space for recruitment, helps explain the relatively greater success of bryozoans under corals, despite intensive overgrowth competition with sponges. Moreover, larval recruitment by bryozoans, although higher than for sponges, is much lower than in temperate waters. This is correlated with extremely low food availability for suspension feeders on most island reefs. The few larvae that do recruit are usually overgrown by clonal encroachment. Thus, larval recruitment plays a minor role compared to biological interactions in determining community composition under corals.

7.4 Settlement panels

Artificial substrata have been used extensively to investigate factors controlling epibenthic community structure. Unfortunately, however, results of most of these studies are difficult to interpret because artificial substrata rarely resemble natural substrata in the local environment. For example, most settlement panels are as small as shells or stones but are fixed rigidly to supporting frames so that they cannot move about (be disturbed) like natural substrata. Another serious problem is that panels are commonly suspended from piers or floats, isolated from predators they would normally encounter at the bottom, and from sources of larvae in the benthic population. Because most bryozoan (and other clonal animal) larvae travel very short distances (§ 5.5.1), the relative composition of colonizers onto these panels may be drastically different from values in nearby natural environments. Even the size of panels can be important, because larvae of many species appear to exhibit strong size preferences at settlement (Jackson 1977b, Keough 1984a). Not surprisingly, communities on such artificially secured, isolated, and odd-sized substrata are often grossly different from those on nearby natural substrata. For example, virtually all reported counter examples to the general pattern of dominance of subtidal hard substrata by clonal animals or algae (Jackson 1977a) are based on settlement panels (Castric 1974, Haderlie 1974, Schoener & Schoener 1981, Greene & Schoener 1982, Greene *et al.* 1982, Rubin 1985).

Settlement studies can be useful if care is taken to mimic natural conditions. Keough (1984a, b) used panels similar to *Pinna bicolor* in size, orientation, and position on the bottom to investigate the dynamics of this bivalve's epibiota. Winston & Jackson (1984) used panels similar in size, orientation, and position to natural coral substrata to investigate recruitment dynamics of cryptic coral reef communities. In both cases, the rigid positioning of the panels and the communities that developed on them closely resembled the natural situations. Likewise, pier pilings resemble boulders or rock walls in their spatial scale, relation to the bottom, and permanence.

7.5 Characteristics of abundant encrusting bryozoans on stable substrata

The abundance and persistence of bryozoans encrusting stable substrata depend on many factors including their size, geometry, growth potential, and possession of specific interactive mechanisms. In general, many such factors operate simultaneously, thereby greatly complicating the nature of biological interactions and decreasing the predictability of their outcomes (Jackson & Buss 1975, Jackson 1979b, Russ 1982, Buss 1985).

Colony size
Larger colonies survive longer than smaller colonies (§ 5.4). Larger colonies win relatively more overgrowth interactions than smaller colonies (Day 1977, Buss 1980a, Russ 1982, Winston & Jackson 1984). They also are more likely to survive partial predation or physical damage, and to regenerate injuries more rapidly than smaller colonies (Jackson & Palumbi 1979, Jackson *et al.* in preparation). Larger colonies presumably have greater resources to deal with localized threats or injuries. Colony size is also correlated with other important factors such as colony thickness, which also confers advantages.

Colony thickness
Thicker colonies are more resistant to overgrowth than thinner colonies and are better able to overgrow their neighbors (Jackson & Buss 1975, Osman 1977, Buss 1980a, 1981b, Rubin 1982). Thicker colonies may also be more

resistant to grazing predators and faster to regenerate injuries than thinner colonies (Jackson & Hughes 1985, Jackson *et al.* in preparation). Thickness increases with height of zooids and through various forms of frontal budding (§ 3.1.2). Frontal budding may occur in a regular pattern of colony development or may be stimulated locally at points of interaction or obstruction.

Growth rate
Obviously, faster growing colonies can pre-empt space more quickly than slower growing colonies. Growth rate is also positively correlated with success in overgrowth interactions and rates of regeneration of injuries (Buss 1981a, b, Palumbi & Jackson 1982, 1983, Winston & Jackson 1984, Jackson *et al.* in preparation).

Growth flexibility
Encrusting colonies that can raise their colony margin away from the substratum are favored in overgrowth interactions against colonies that cannot (Fig. 7.7a, Jackson *et al.* 1971, Gordon 1972, Stebbing 1973b, Jackson & Buss 1975, Harmelin 1976, Jackson 1979b, Rubin 1982). Similarly, colonies that can form new growing edges in regions of interaction may halt overgrowths by formation of vertical barriers or even reverse their outcome (Fig. 7.7b, Jackson 1979b, 1983). Another aspect of flexibility is the ability of a colony to change its direction of growth in anticipation of interactions with other colonies (Buss 1981b). Change of direction is facilitated by localized production of kenozooids, and may be stimulated by sensation of feeding currents of approaching colonies. This is important because the outcome of overgrowth interactions is highly dependent on the angle at which colonies encounter each other (Jackson 1979b, Buss 1981b, Rubin 1982).

Zooidal morphology
The design, thickness, and composition of the frontal and lateral walls of zooids determine their resistance to different kinds of forces, and, by analogy, to certain types of predators (Best & Winston 1984). All abundant species on exposed reef surfaces are very heavily calcified (Best & Winston 1984, Jackson & Hughes 1985, Ristedt & Schuhmacher 1985). The presence of a well-defined cryptocyst seems to limit feeding by small predators such as isopods that consume the contents of individual zooids (Buss & Iverson 1981).

Specialized defenses
Formation of stolonal outgrowths (Fig. 7.8) of zooids at the growing margin of colonies may affect the outcome of competitive encounters by making the colony thicker, and therefore harder to overgrow, or by actively interfering with the functioning of peripheral zooids of neighboring colonies (Gordon

Figure 7.7 Morphology of aggression in cheilostome overgrowth interactions. (a) *Stylopoma spongites* (top) overgrowing *Steginoporella* sp. (bottom) in a flank attack on the right while *Steginoporella* has raised its growing edge to stave off overgrowth in the frontal encounter to the left; (b) *Parasmittina* sp. (top) has halted overgrowth by *Reptadeonella costulata* (bottom) by formation of a new wall-like growing edge along the line of contact between them. (Field of view approximately 1 cm; both from Jackson 1983, courtesy of Plenum Press.)

1972, Osborne 1984). However, the degree to which stolonal outgrowths may be specialized defensive or aggressive structures has not been investigated. Zooidal spines can effectively reduce rates of predation, and their presence can be induced experimentally by introduction of predators (Yoshioka 1982b, Harvell 1984a). Spines may also slow or prevent overgrowth, but this has not been demonstrated experimentally (Stebbing

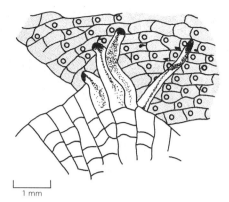

Figure 7.8 Formation of stolonal outgrowths by *Stylopoma duboisii* in process of overgrowing *Thalamoporella tubifera*. (After Osborne 1984, courtesy of *Australian Journal of Marine and Freshwater Research*.)

1973b). Accumulation of waste products may inhibit predators (Harvell 1984b), but there is no evidence for the production of specific chemical defenses by encrusting species.

Feeding interference

Interference competition for food has been demonstrated in the laboratory for two species of anascans that encrust cobbles in Panama (Buss 1980b). In the case examined, the species with the larger lophophores, *Onychocella alula*, produced stronger feeding currents which disrupted the feeding currents of the peripheral zooids of the species with smaller lophophores, *Antropora tincta* (Fig. 7.9). In addition, suspended food may be greatly reduced in some benthic environments (Buss & Jackson 1981), which may be a primary determinant of contrasting distributions of different bryozoan species or major taxa, such as the predominance of cyclostomes versus cheilostomes in cavities with restricted circulation (Harmelin 1977, 1985).

Mutualism

Several species of abundant encrusting cheilostomes are colonized by the stoloniferous hydroid *Zanclea* (Hastings 1930, Osman & Haugsness 1981, Ristedt & Schuhmacher 1985). The presence of *Zanclea* on colonies of *Celleporaria brunea* was demonstrated to increase the latter's success in overgrowth interactions with other bryozoans and to decrease densities of predatory flatworms as compared to colonies lacking *Zanclea* (Osman & Haugsness 1981). In return, host bryozoans commonly form calcareous structures over the hydroids' stolons which may provide protection, as well as preservable structures which should allow study of the evolution of this interaction using fossils.

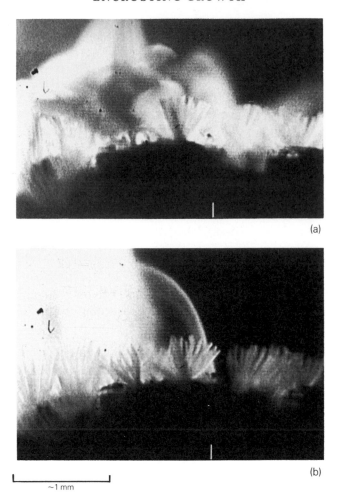

(a)

(b)

└──────────────────┘
    ~1 mm

Figure 7.9 Feeding interference between *Onychocella alula* (left of bar) and *Antropora tincta* (right of bar). (a) Suspension of milk (white blur) was introduced above colonies; (b) within 19 seconds, feeding by *O. alula* had drawn all the milky water from above the adjacent portion of *A. tincta*; unfiltered milk particles escaping from the margin of *O. alula* travel upward in a plume and are drawn back over *O. alula*. (From Buss 1980b, courtesy of Macmillan Journals Ltd.)

Table 7.2 lists the most abundant encrusting bryozoan species on a variety of substrata ranging from ephemeral algal fronds to long-lived reef corals. Cheilostomes are clearly the most abundant bryozoans on all these substrata, and are the only abundant species on substrata more stable than dead shells or cobbles. Moreover, among the cheilostomes, ability to grow large, thick, and fast, and to respond in a versatile manner to localized dangers, is strongly correlated with one or more distinctive modes of budding of zooids (Lidgard & Jackson 1982, Lidgard 1985a, b). These are the presence or absence of multizooidal (giant) budding (Figs. 1.21, 1.24) –

which is associated with rapid lateral extension and large colony size – and self-overgrowth or frontal budding – which are associated with increased colony thickness. These budding modes also contribute to increased flexibility of growth. Potential for multizooidal budding, self-overgrowth, and frontal budding is indicated for each cheilostome species listed in Table 7.2. Species are scored from 0 to 3, depending on the numbers of these traits exhibited.

The resulting pattern is striking. Ephemeral, unstable substrata are dominated by species with low budding scores, stable substrata by species with high scores. As we have seen, persistence on large stable substrata involves more-or-less continual interactions with highly aggressive spatial competitors and predators. Coping with these problems in turn depends upon flexibility conferred by the budding characteristics enumerated in Table 7.2. Since these budding patterns are polyphyletic, their prevalence on stable substrata is almost certainly adaptive. This view is supported by the striking and progressive increase in the proportion of encrusting cheilostome species with zooidal (includes multizooidal) and frontal budding throughout cheilostome history (Fig. 7.10, Lidgard 1986).

7.6   Paleoecology of encrusting bryozoans

Bryozoans have encrusted plants, shells, and corals since the Middle Ordovician, and in many cases have been the most abundant group, especially on the less stable substrata. Several ancient encrusting communities have been described, but too little ecological information is available for confident assessment of factors that controlled their distributions.

7.6.1   Fossil bryozoans encrusting plants

The best fossil data are from the Masstrichtian of the Netherlands, where exceptional preservation clearly demonstrates extensive plant–bryozoan relationships (Voigt 1973, 1979, 1981). Many species encrusted algal fronds and seagrass blades, and some appear to have been restricted to these ephemeral substrata (Fig. 7.11). Included among these are several uniserial bryozoans, hydroids, and serpulids whose unique mode of preservation (molds in the basal encrusting surfaces of multiserial bryozoans) demonstrates that overgrowth was common in these communities, and was similar in outcome to such interactions today.

More stable plant substrata, particularly the stems of the abundant seagrass *Thalassocharis bosqueti*, supported a more diverse and abundant bryozoan fauna, dominated by rigidly erect species (Fig. 4.21). As already noted in Section 4.5.2, this suggests considerably less predation by durophagous fishes and sea urchins than occurs in such shallow environments today, which are dominated by encrusting or flexible erect forms (Harmelin 1976). Nevertheless, encrusting bryozoans were also abundant

Table 7.2 Budding characteristics of the most abundant (per cent cover) species of encrusting bryozoans on ephemeral and stable substrata. * = ctenostome,† = cyclostome, the remainder are cheilostomes. Budding patterns are indicated only for cheilostomes, as explained in the text.

Substratum	Species	Multi zooidal budding	Self-over growth	Frontal budding	Total score
Macrocystis fronds (1)	*Membranipora membranacea*	+	−	−	1
	Celleporella hyalina	−	−	+	1
	Lichenopora buskiana†	‡			
Laminaria fronds (2)	*Electra pilosa*	−	−	−	0
	M. membranacea	+	−	−	1
	C. hyalina	−	−	−	1
Fucus fronds (3)	*Alcyonidium hirsutum**				
	*A. polyoum**				
	*Flustrellidra hispida**				
	E. pilosa	−	−	−	0
	M. membranacea	+	−	−	1
	C. hyalina	−	−	+	1
Rhodymenia fronds (4)	*C. hyalina*	−	−	+	1
	Microporella cribosa	−	−	−	0
	Thalamoporella californica	+	−	−	1
	Fasciculipora pacifica†				
Homarus americanus (5)	*Alcyonidium polyoum**				
dead shells and small stones (6)	*Fenestrulina malusi*	−	−	−	0
	Microporella ciliata	−	−	−	0
	Escharella immersa	−	−	−	0
dead shells (7)	*Parasmittina* sp.	−	−	−	0
	Smittina sp.	−	−	−	0
	Rhynchozoon sp.	−	−	−	0
	Reptadeonella sp.	+	−	−	1

Table 7.2   (cont.)

Substratum	Species	Multi zooidal budding	Self- over growth	Frontal budding	Total score
	Stylopoma spongites	+	+	+	3
Aequipecten gibbus (8)	*Schizoporella unicornis*	+	+	+	3
Gryphus vitreus (9)	*Schizomavella auriculata*	+	+	+	3
	Escharina hyndmanni	−	−	−	0
cobbles (10)	*Crassimarginatella papulifera*	−	−	−	0
	Micropora mortenseni	−	−	−	0
cobbles & boulders (11)	*Schizoporella errata*	+	+	+	3
cobbles & boulders (12)	*Antropora tincta*	−	+	+	2
	Onychocella alula	−	−	−	0
Pinna nobilis (13)	*Schizoporella sanguinea*	+	+	I	3
	Schizoporella sp.	+	+	+	3
Pinna bicolor & pier pilings (14)	*Schizoporella schizostoma*	+	+	+	3
	C. fusca	+	I	+	3
	P. ruigii	I	+	+	3
pier pilings (15)	*S. errata*	+	I	+	3
reef corals & large coral rubble (16)	*Steginoporella* sp.	+	+	−	2
	Steginoporella magnilabris	+	+	−	2
	Reptadeonella bipartita	+	+	−	2
	R. costulata	+	+	−	2
	S. spongites	+	+	+	3
	Parasmittina sp.	+	+	+	3
	Petraliella bisinuata	+	+	−	2

Table 7.2 (cont.)

Substratum	Species	Multi zooidal budding	Self- over growth	Frontal budding	Total score
corals, coral rubble, & reef frame work (17)	*Trematooecia aviculifera*	+	+	+	3
corals, coral rubble, & reef frame work (18)	*Rhynchozoon* sp.	+	+	+	3
corals, coral rubble, & reef frame work (19)	*Rhynchozoon llarreyi*	+	+	+	3

‡Budding by *Lichenopora* produces a common multizooidal margin similar in its extensive flexibility of growth to multizooidal cheilostomes (Stebbing 1973b).

Study sites and references: (1) southern California, Bernstein & Jung 1979, Yoshioka 1982a; (2) British Isles, Seed & Harris 1980; (3) British Isles, Stebbing 1973a, Boaden *et al.* 1975, O'Connor *et al.* 1980, Wood & Seed 1980; (4) southern California, Bernstein & Jung 1979; (5) Connecticut, Dexter 1955; (6) Irish Sea, Eggleston 1972a; (7) Venezuela, Gleason & Jackson, in preparation; (8) North Carolina, Wells *et al.* 1964; (9) Mediterranean, d'Hondt 1984; (10) New Zealand, Ryland 1975; (11) Massachusetts, Osman 1977; (12) Panama, Buss 1980a & 1981b, LaBarbera 1985; (13) Adriatic Sea, Zavodnik 1967; (14) South Australia, Kay & Keough 1981, Keough 1983, 1984a,b; (15) North Carolina, Karlson 1978; (16) Caribbean basin, Jackson 1983, Winston 1984a, Gleason & Jackson in preparation; (17) Jamaica and Panama, Jackson & Hughes 1985, Jackson unpublished; (18) Pacific Panama, Jackson unpublished; (19) Red Sea, Ristedt & Schuhmacher 1985.

on Maastrichtian seagrass stems, and included both multilaminate and unilaminate growth forms (Fig. 7.11). Overgrowths involving these species are evident in a few of Voigt's illustrations, but neither their frequency nor that of regenerative budding (Fig. 5.16) have been determined. Given the exceptional preservation of this community, however, it is certain that no species dominated the stems by extensive multilaminate growth comparable to that of Pleistocene to Recent *Schizoporella floridana* on the seagrass *Thalassia* (Fig. 7.12). Massive multilaminate growths of this species are up to 46 laminae thick and comprise an extensive bryozoan facies in which the rock mass locally exceeds 70% *Schizoporella*. The *Schizoporella* facies blankets more than 5,000 km^2 of southern Florida and exists in vast areas of comparable environments in the Bahamas today (Hoffmeister *et al.* 1967). This species, like the *Schizoporella* listed in Table 7.2, displays multizooidal and frontal budding and self-overgrowth.

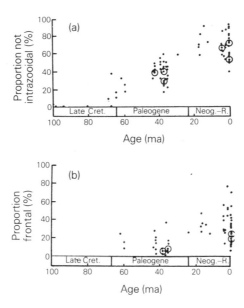

Figure 7.10   Proportion through time of species with different budding patterns within North American faunas of encrusting cheilostomes: (a) zooidal budding (intrazooidal budding is the inverse), (b) frontal budding. (Based on data from Scott Lidgard.)

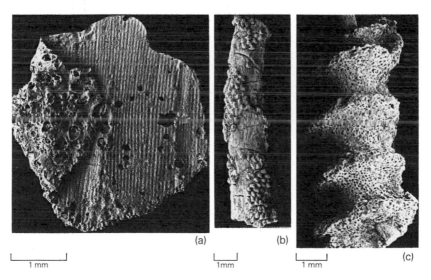

Figure 7.11   Association of encrusting bryozoans and seagrass from Maastrichtian of the Netherlands: (a) leaf of seagrass molded in remains of base of encrusting cheilostome seen at left, (b) several zoaria of *Kunradina bicincta* on a seagrass stem, (c) spiral growth of *Coelochlea torquata* encrusting a stem. (From Voigt 1981, courtesy of Olsen & Olsen.)

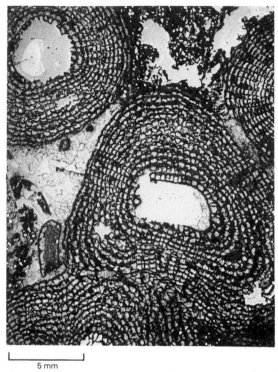

|_____|
        5 mm

Figure 7.12 Transverse section of *Schizoporella floridana* from Miami Limestone (Pleistocene) showing once hollow tubes where the colony encrusted rhizomes of seagrass. (Photograph courtesy of A. H. Cheetham.)

7.6.2   *Fossil bryozoans encrusting dead shells and skeletons*

The best information is for epizoans encrusting the overturned calyces of two species of camerate crinoids (*Eucalyptocrinites crassus* and *E. magnus*) in the Silurian Waldron Shale, Tennessee (Fig. 7.13, Liddell & Brett 1982). The calyces provided abundant (up to 13 per m^2 on bedding surfaces), small (2–4 cm diameter) substrata for at least 27 encrusting species. Bryozoans were the most diverse group (16 species) and covered the most space (12% and 28% cover on *E. crassus* and *E. magnus* respectively). No other fossilized group occupies more than 1% space, and bare space is extensive (71% and 86%). Tubiculous worms, mostly *Spirorbis*, occur on most of the calyces; echinoderm holdfasts, brachiopods, and borings on far fewer. *Saginella elegans* (= *Berenicea consimilis*) is the most abundant bryozoan; it occurs on roughly half the calyces of both species and comprises 76% of the total bryozoan cover. Despite the extensive bare space, overgrowths of fossils are common (Fig. 7.13) and were apparently contemporaneous, as evidenced by the equally good preservation of most colonies and, more importantly, by the local reversal in outcome of interactions between the

Figure 7.13   Calyx of *Eucalyptocrinites* heavily encrusted by bryozoans: (a) majority of surface covered by *Saginella elegans*, (b) overgrowth of one ceramoporoid by another and both over *Saginella*, (c) ceramoporoid over *Saginella*, note raised and thickened edge. (From Liddell & Brett 1982, courtesy of the Paleontological Society.)

same colonies. Outcome of 225 interspecific interactions involving unworn bryozoan specimens were compiled. The most interesting result is that the most abundant species, *S. elegans*, was by far the most frequently overgrown. This suggests that abundances of epizoans were determined more by rates of recruitment and growth than by competition, and that the apparently low cover by epizoans may be accurate, reflecting an ancient balance between these processes and disturbance or destruction of the calyces. Comparable faunas on small skeletal substrata occur throughout the fossil record (Ager 1961, Spjeldnaes 1975, Gundrum 1979, Taylor 1979b, 1984, Taylor *et al.* 1981). Overgrowths of fossils are usually evident, as are signs of grazing by snails and echinoids. Apparent (preserved) cover is low in all cases.

7.6.3   Hardgrounds and reefs

There have been numerous studies of fossil hardgrounds (reviewed in Palmer 1982), but no good ecological investigation of modern equivalents for comparison. Bryozoans were common inhabitants of upper surfaces of hardgrounds during the Paleozoic, but became restricted to cavities and overhangs during the Mesozoic, as they still are today. Sediment scour and other forms of physical stress are unlikely to have changed radically during the Mesozoic. Thus these switches in distributions must reflect differing impacts of biological interactions. The most likely scenario to date is that increased abundance and efficiency of durophagous predators, coupled with competition by newly resistant forms of prey, resulted in displacement of less resistant sessile taxa (cf. Steneck 1983). Paleozoic reefal and carbonate bank deposits commonly include remains of once dense populations of encrusting as well as erect bryozoans that lived on exposed (upward-facing) surfaces (Ch. 4, Ross 1970, Walker & Ferrigno 1973, Kapp 1975, Wilson 1975, Cuffey 1977), whereas such forms are now virtually restricted to vertical walls, cryptic habitats, and rubble. Again, escalation in the inventions of new kinds of apparatus (teeth, jaws, armor) important in biological interactions seem the most likely explanation.

Appeal to changes in biological interactions to explain historical changes in bryozoan distributions does not imply that ecological processes are somehow fundamentally different. There have been predators and overgrowths on hard substrata since the Cambrian (review in Jackson 1983). Rather, the weapons have changed and so also the kinds of organisms best able to defend against them. Nothing more can be said, however, until impacts of different causes of mortality of bryozoans have been carefully documented for many fossil populations. This will require detailed, quantitative analysis of amounts of injury, overgrowth, and regeneration in hundreds of colonies from each assemblage. By far the most deserving material for such investigation is the epifauna of shells and skeletal debris where encrusting bryozoans have most prospered since the end of the Ordovician. Documentation of amounts and causes of mortality on these substrata should reveal whether or not a biological "revolution" (Vermeij 1977, Steneck 1983) did indeed occur on these substrata, and its relation, if any, to the progressive replacement of cyclostomes by cheilostomes during the Cretaceous and early Paleogene. The progressive restriction of cyclostomes to more cryptic habitats than cheilostomes (Jackson et al. 1971, Harmelin 1976, 1977, Harmelin et al. 1985) strongly suggests a causal relationship.

8 Erect growth: problems of breakage and flow

Three-dimensional growth increases the surface area and volume of a sessile animal relative to that of the substratum on which it lives. The greatest increase is achieved by erect growth. The simplest erect form, a cylinder or sheet attached at one end, increases its surface area in direct proportion to height (Fig. 8.1a, d), while the area of the attachment increases minimally or not at all. Branching of an erect colony increases surface area exponentially with upward growth, depending on the frequency of branch division (Fig. 8.1b, c) and on branch diameter. While area of attachment generally remains small, the area of substratum that is overtopped or "shaded" by branched colonies increases with colony growth. In all cases, however, the area of substratum shaded is less than the surface area of the erect colony. Erect growth is therefore a commitment to life within the water mass and a retreat from the substratum. Among other possible problems, erect growth generates dependence on integrity of supporting parts of the colony or, alternatively, the ability to turn colony breakage to benefit.

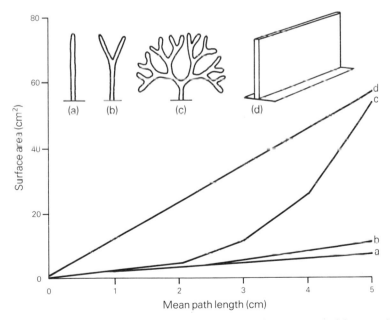

Figure 8.1 Surface areas at 4 cm height of (a) erect, unbranched cylinder, (b) erect cylinder branched at 2.5 cm, (c) erect cylinder branched at 1.0 cm intervals, each with 0.5 cm diameter, and (d) erect sheet 5 cm × 0.2 cm.

8.1   Consequences of erect growth

Erect growth has been postulated to confer three principal advantages over encrusting growth (Cheetham 1971, Jackson 1979a):

(a) high tissue area and volume which increase feeding and reproductive capacity per unit area of substratum;
(b) increased access to food in the water column; and
(c) greater isolation from competitors, predators, and sediments on the substratum.

Much circumstantial evidence supports these ideas, but none of them has been tested experimentally.

There is no doubt that tissue area is increased relative to substratum area in erect bryozoans, and, thus, that numbers of zooids, lophophores, and gonads also increase. However, the benefit of increased surface area may be partially offset if only a part of the area of the colony is able to feed and reproduce. Some parts of erect colonies may have access only to previously filtered water, lack feeding polypides, and serve primarily to support the more active portions. For example, only branch tips bear functioning polypides in the flexible, erect colonies of *Epistomaria bursaria* (Dyrynda 1981, Dyrynda & King 1982). Even in this case, however, surface area is greater than for encrusting growth because of the generally exponential increase in number of branch tips, and therefore functional feeding zooids, with growth. Similarly, zooids capable of reproduction may be limited to certain regions of a colony (Correa 1948), as in species of *Bugula* (Fig. 8.2).

Evidence for postulated advantages and disadvantages of erect growth in biological interactions comes from studies of overgrowth competition, larval recruitment, and predation. Erect growth may allow relative escape from overgrowth by potentially superior competitors. For example, the compound ascidian *Botryllus* typically overgrows all other encrusting organisms it encounters, including bryozoans, but it grows around the base of erect *Bugula* colonies, leaving the protruding parts unaffected (Grosberg 1981). Moreover, larvae of encrusting species avoid settling near *Botryllus*, whereas larvae of *Bugula* do not.

Botryllus is soft-bodied and cannot push over *Bugula* as it grows over the substratum. In contrast, the calcified growing edge of the encrusting cheilostome *Schizoporella* can push over and overgrow erect *Bugula*, except when the latter occur in dense clumps of many colonies. This group advantage in resisting overgrowth is the probable basis of strongly gregarious larval settlement by *Bugula turrita*, which occurs in spite of decreased growth of clumped, as compared to solitary, colonies of this species (Buss 1981a).

As described earlier (§ 4.6.1), erect bryozoans are more vulnerable to grazing than are encrusting forms. However, extensive studies of biological interactions involving erect bryozoans have been partially hampered by

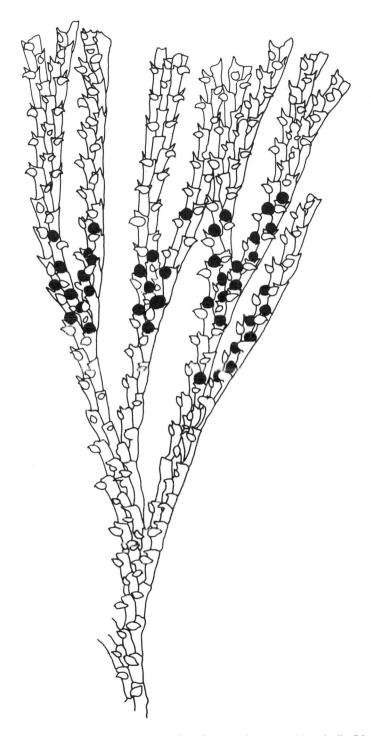

Figure 8.2 Portion of a branch system from *Bugula turbinata*, with ovicells (blackened) concentrated in a zone proximal to branch tips.

poor access to live material of most erect types, the majority of which live in relatively deep water and tend to occur in remote places like the Antarctic (e.g. Winston 1983). Faced with these difficulties, and buoyed by the tendency of paleontologists to conceptualize erect growth forms as a series of biomechanical engineering solutions to physical problems, the focus in the study of erect bryozoans has been almost exclusively in relation to potential breakage or damage by water movement. This approach has led to exciting demonstrations of convergence and of progressive trends in biomechanical design. However, cause and effect have not been demonstrated by experimental manipulation, although they are often reasonably inferred on the basis of mechanically consistent interpretations.

Accordingly, in this chapter we emphasize some of the more outstanding advances in understanding of the physical and mechanical properties and the architecture of erect bryozoans, and their possible adaptive significance. In so doing, we make a fundamental division of erect forms into flexible and rigid colonies. This dichotomy clearly relates to investment in persistence, the former generally being more ephemeral than the latter. Among rigid forms we consider separately colonies with different branch morphologies (unilaminate, bilaminate, and radial).

Some biomechanics and mechanical terms are inescapable. Forces experienced by any erect benthic organism may be of two types. **Concentrated loads** are localized and may be due to impact of debris or mobile organisms, whereas **uniform loads** are dispersed across large regions, such as drag due to flow of water. Either of these types of load bends colonies that have a single basal trunk as if they were a singly supported beam. The **bending moment (Mb)**, which is a measure of intensity of a bending force, is determined in such conditions by

$$\mathbf{Mb} = \mathbf{F} \cdot d$$

where **F** is the deforming force and d is distance along the structure from the point of application of the deforming force. Therefore, bending moment from any force applied to an erect colony increases towards and reaches a maximum at the lowest point, which may be some point of flexure in flexible colonies and is the colony base in rigid colonies.

Bending moment due to uniform load (**drag**) increases in magnitude with growth of a colony for two reasons. First, as the colony grows higher, the center of the drag force also moves higher and thereby increases d. Secondly, as the colony grows, its cross sectional area increases exponentially and therefore so does the drag component of **F**. Branching properties of colonies determine the rate at which total cross sectional area grows relative to height, and consequently are important in determining bending moment that results from drag.

At any point along an erect bryozoan colony, **stress** (σ) is directly proportional to bending moment and inversely proportional to a factor designated as the **section modulus** (Z),

$$\sigma = \frac{Mb}{Z} = \frac{F \cdot d}{Z}$$

Z is determined in large part by branch thickness or radius. In colonies in which there is no proximal branch thickening, there is also no change in Z, whereas **Mb** continues to increase proximally. Therefore, stress due to drag forces builds to a highly elevated maximum at the colony base (Fig. 8.13, curve A). In contrast, sufficient increase in Z, such as would be caused by very substantial branch thickening, may result in decreased stress toward the colony base.

Resistance to bending of rigid materials such as calcified bryozoan skeletons is termed **modulus of rupture** (*Mr*). It is the stress at the point of rupture, calculated as though all the deformation were elastic, and is measured in units of stress per cross sectional area.

Further definitions and explanations of the mechanical terms used here, as well as much more extended discussions of biomechanics, may be found in Wainwright *et al.* (1976), Vogel (1981), and Cheetham & Thomsen (1981). A painless and entertaining introduction to mechanics is provided by Gordon (1978).

8.2   Flexible erect growth

Most erect bryozoans found in regions of vigorous water motion are flexible. Consequently, they can accommodate most of the stress upon them by elastic deformation. Flexibility may result from absence of calcification, light calcification, or presence of noncalcified nodes (joints) between strongly calcified branch segments. All the basic branch morphologies, including radial, bilaminate, and unilaminate forms, have flexible representatives.

Colonies that have noncalcified skeletons, such as *Alcyonidium* with radial branches, and those that have lightly calcified skeletons, such as *Bugula* with unilaminate and *Flustra* with bilaminate branches, may be flexible throughout. Alternatively, noncalcified or lightly calcified colonies may have relatively robust branch segments that alternate with short, severely narrowed regions where the bending moment is accommodated by the high tensile strength of the cuticular skeleton and increased flexibility at the narrowed region. For example, *Spiralaria florea* has tightly twisted bilaminate branches with narrow points of origin from parent branches (McKinney & Wass 1981). If the twisted bilaminate branches are modelled as structural cylinders, which they resemble, and using typical segment diameters as shown in Fig. 8.3b, stress at the narrow base of a branch segment can be estimated as

$$\sigma = \frac{4 \cdot Mb \cdot r}{r^4} = \frac{4 \cdot Mb \, (0.15 \, mm)}{(0.15 \, mm)^4} = 1185Mb$$

where **Mb** = bending moment and r = radius. If the base of the branch segment had the same diameter as the midportion, stress at the base could be estimated as

$$\sigma = \frac{4 \cdot \mathbf{Mb} \,(0.75\,\mathrm{mm})}{(0.75\,\mathrm{mm})^4} = 9.5\mathbf{Mb}$$

The increase in stress at the base of a branch segment due to the decreased diameter is therefore about 10^2. The stress is relieved through bending and stretching of the flexible cuticle, which inhibits its buildup and transfer from the numerous distal branch segments to the branches in the colonies' basal regions.

Strongly calcified colonies may have rigidly calcified branch segments separated by narrow, flexible cuticular nodes. Such articulation evolved in numerous erect bryozoans with cylindrical branches (arthrostylid crypto- stomes and cellariid cheilostomes), with unilaminate branches (some arthrostylid cryptostomes, crisiid cyclostomes, scrupocellariid cheilo- stomes), and with bilaminate branches (some ptilodictyine cryptostomes). In at least some cheilostomes, each node consists of a single, diminutive, noncalcified kenozooid (Cook 1975), and in others, of the noncalcified proximal ends of otherwise unmodified autozooids (Harmer 1923, 1926). The cuticular nodes serve the same functions as the branch bases in *Spiralaria*; they are points of concentration and accommodation of bending stress by the tough cuticle (Fig. 8.3a). Reduction in branch diameter from 1.1 mm in the calcified segments to 0.3 mm in the noncalcified nodes of *Tropidozoum* (as in Figure 8.3a) results in about 50 times more stress at the nodes than if the larger diameter were retained there.

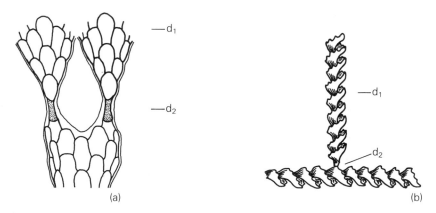

(a) (b)

Figure 8.3 Increase in stress at narrowed articulation points in (a) well-calcified *Tropidozoum* with uncalcified kenozooids at articulation points (from Cook 1975, courtesy of the Université Claud Bernard, Lyon); and (b) lightly calcified *Spiralaria florea*. $d_1 = 1.1$ mm in (a), 1.5 mm in (b); $d_2 = 300\,\mu$m in (a) and (b).

8.3 Rigidly erect growth

Rigidly erect growth is accomplished by calcification of the body wall, either within zooids or extrazooidally, extending continuously from base to growing tips. The process of calcification apparently involves both energetic costs (production and mobilization of skeletal materials) and temporal costs (Palmer 1981), which are absent or less for flexible colonies. Temporal costs result from maximal tissue growth being set by the maximum *rate* at which skeleton may be produced rather than by the energy in the food that is available for growth. This limitation may result in lesser total surface area, and therefore a lesser ability to utilize space and food resources, which would cause an increasingly lowered growth rate with increased size, relative to flexible colonies, although some moderately well calcified erect colonies can extend at least $0.5 \, mm \, day^{-1}$ (Fig. 8.4). Presumably, growth away from the substratum can be more rapid in species in which only the distal extremities are kept alive, rather than diverting nutrients to nonfeeding proximal parts where deposition of skeleton would augment the proximal section modulus.

8.3.1 Rigidly erect colonies: unilaminate

Narrow branches, rarely wider than 2 mm, characterize erect unilaminate colonies (Fig. 6.15c). The slimness of the branches appears to be a means of

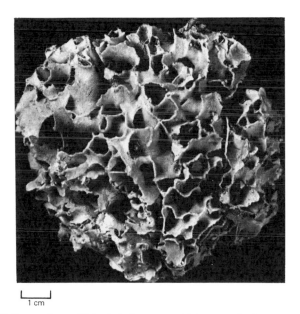

L___J
1 cm

Figure 8.4 Bifoliate colony of *Membranipora savartii* that grew in three months (December 25, 1982–March 25, 1983) at 5 m depth in Dona Paula Bay, Goa, India. (Courtesy of G. Hillmer.)

minimizing interference with passage of water filtered by the lophophores that overlie one surface. In flexible unilaminate colonies the narrow branches can shed great amounts of stress by elastic deformation, but rigidly calcified thin branches may be easily broken with even fairly low stress. Ease of breakage is in part offset by lateral branch linkage, which reduces the length along which bending moment must be absorbed by a single branch, because part can be transferred to nearby branches as compression or tension at branch junctions (Fig. 8.5, Cheetham 1971). Costs of linkage include uneven distribution of feeding lophophores on sinuous, anastomosed branches and, in others, secretion of more carbonate skeleton per feeding zooid in order to form links between straight branches.

One would expect that among rigidly erect unilaminate colonies, lateral branch linkage would be more advantageous in the presence of strong water motion, and unlinked branches would be restricted to areas of weak water movement. This is indeed the case. In the Mediterranean, colonies of *Sertella*, which form net-like colonies of linked branches, can occur in shallow channels with vigorous flow (Hass 1948), while taxa such as *Hornera frondiculata* and *Schizoretepora solanderia*, with unlinked rigid unilaminate branches, are restricted to quieter bottoms (Lagaaij & Gautier 1965). A similar pattern exists among several genera of Upper Mississippian fenestrate bryozoans distributed through various carbonate lithologies that are inferred to represent a range of open-shelf, near-shoal, and back-barrier environments. *Penniretepora*, the only genus with nonlinked branches among the group, is restricted to mud-dominated deposits that represent quieter-water areas (Fig. 8.6, McKinney & Gault 1980). Wherever they co-occur, colonies with linked branches are consistently much larger than,

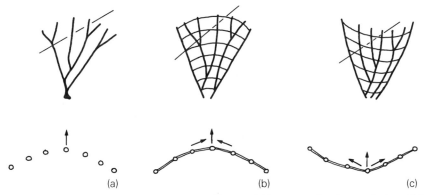

(a)   (b)   (c)

Figure 8.5 Schematic diagram of absorption of forces within erect unilaminate colonies. (a) In unlinked branches the bending moment is confined to a single branch and its directly supporting segments, and thus continues to build to the base of the colony. (b) Force directed into the concave surface of a net composed of linked branches·is translated by linkages to adjacent branches as tension where the branch originally involved pulls on the links. (c) Force directed into the convex surface of a net composed of linked branches is translated by linkages to adjacent branches as compression where the branch originally involved pushes against the links.

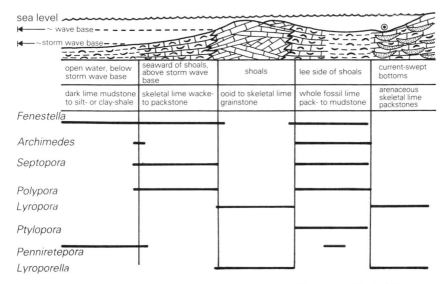

Figure 8.6 Distribution of erect unilaminate bryozoans in Chesterian (Viséan–Namurian) deposits of eastern North America. Bull's-eye represents shore-parallel current. (After information in McKinney & Gault 1980.)

and numerically dominate over, those with nonlinked branches. The ability to grow larger may have been a direct result of increased strength provided by branch linkage, which permitted colonies to absorb more stress, or it may relate in part to control of velocity of ambient flow through the colony (§ 8.3.3).

Linked branches seem the more derived character state in comparison with nonlinked branches. This is because patterned linkage of unilaminate branches to form a porous sheet requires growth of branches in a plane, coordinated response of adjacent branches so that the linkages occur, and growth of branch tips in a continuous zone. Such requirements are not as stringent for nonlinked unilaminate branches. Moreover, the geologic distribution of branch linkage in all the major unilaminate clades is consistent with interpreting it as the more derived state (Fig. 8.7). In each case, branch linkage, and the more highly ordered states of linkage, either follow after or appear stratigraphically with the unlinked inferred ancestral state.

The most successful erect unilaminate bryozoans were Paleozoic fenestellid fenestrates that had straight branches connected by skeletal bars. Many of them had such highly coordinated branch widths, interbranch spacing, and spacing of zooidal orifices along branches that a rhombic pattern of orifice placement was produced (Fig. 6.19). Such an arrangement should have maximized areal cover of lophophores for feeding while minimizing interference between them, a factor that probably contributed to their great abundance.

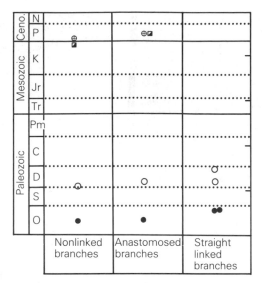

Figure 8.7   Times of appearances of types of branch linkages in some clades of Paleozoic and post-Paleozoic erect unilaminates: (●) fenestellid Fenestrata; (○) pinnate Fenestrata; (▪) hornerid Cyclostomata; (⊕) sertellid Cheilostomata.

Fenestrate bryozoans lived predominantly in moderately energetic to relatively quiet environments or microenvironments, either near or below normal wavebase or in sheltered localities (McKinney & Gault 1980 and references therein). Some have thickened and coalesced branches in basal parts and are found as relatively whole colonies (Fig. 8.8). However, most fenestrates, including the characteristic fan- or funnel-shaped forms, are collected only as fragments. Breakage has variously resulted from weathering, post-depositional compaction of argillaceous sediment, and pre-depositional stresses (the latter most apparent where fragmentary remains occur in noncompacted limestones).

Two late Paleozoic fenestrate genera, whose habitat preferences and colony forms were distinctly different, routinely broke at points of structural weakness in their colony designs. One of them, *Archimedes*, was capable of building enormous clonal populations in relatively quiet-water habitats, whereas the other, *Lyroporella* (and other lyre-shaped genera), was adapted to survive in vigorously flowing unidirectional currents.

Archimedes

This genus occurs most abundantly in deposits formed in protected carbonate mud bottoms in shallow water (McKinney 1979, McKinney & Gault 1980). Nevertheless, the upright rigid colony form would inevitably have caused increased stress on the colony base with continued growth, even in the gentlest current. The point of origin of an *Archimedes* spiral (Figs. 8.9, 8.10) is roughly one-fifth to one-tenth the diameter of the higher portions

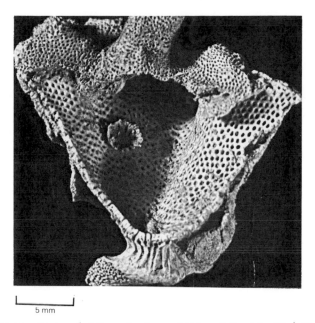

5 mm

Figure 8.8 Unidentified conical Devonian fenestellid from Falls of the Ohio, Kentucky, with secondarily thickened basal attachment. Specimen is encrusted around the outside and across the top by *Fistulipora* sp., and has a young conical fenestellid on the inside of the cone.

of the thickened axis (and as little as one-hundreth the diameter of the entire whorls), so that difference in radius alone must have caused calculated increases in stress of about two orders of magnitude at spiral origins over that in the higher portions. Most *Archimedes* occur as axes from which the radiating branches have broken, or are preserved in such well-lithified limestones that whorl margins cannot be examined. Evidence for fragmentation in *Archimedes* has come primarily from a single locality in an easily decomposed marly limestone where there are a large number of specimens that show secondary spirals originating from whorl peripheries of parent *Archimedes* colonies, as well as other indications of growth from previous colony margins (Fig. 8.9). Most secondary spirals broke from their parent spirals exactly at their point of origin, effectively separating the parent and descendant ramets. Not a single colony base developed from an ancestrula has been discovered at the locality, so it is clear that here, at least, the dominant mode of reproduction in *Archimedes* was by colony fragmentation rather than by larval recruitment (Fig. 8.10), and that relatively few genets are represented by the thousands of colonies, a situation common among branching bryozoans and corals in Recent seas (reviews in Highsmith 1982, Jackson 1986).

Lyre-shaped fenestrates
In a reconnaissance study of the distribution of fenestrate bryozoans in

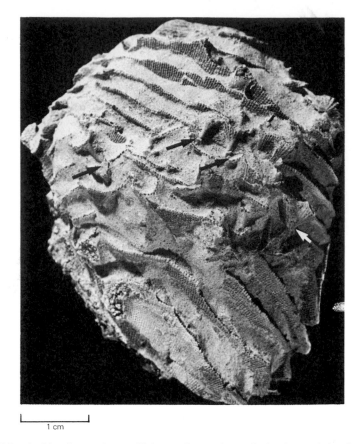

1 cm

Figure 8.9 *Archimedes* specimen with bases of secondary colonies (arrows) developed from the edge of the frond of a larger parent colony – Lower Bangor Limestone (Chesterian; Viséan–Namurian), Fox Trap, Colbert County, Alabama. (From McKinney 1980, courtesy of the Paleontological Society.)

Upper Mississippian rocks of the eastern United States, *Lyropora* (Fig. 6.20) and *Lyroporella* were found only in various calcareous sands, commonly thin quartzose calcarenites or calcareous quartz wackes within finer-grained carbonate sequences. The arenitic beds were interpreted as deposited by unidirectional, perhaps longshore, currents that introduced the quartz sand (Fig. 8.6, McKinney & Gault 1980). *Lyroporella* is similarly limited to calcareous sandstones in slightly older Mississippian rocks in the Grand Canyon, Arizona (Duncan 1969).

Lyre-shaped fenestrates consist of a single, bowed fan, with the proximal and lateral edges of the colony generally forming a smooth curve that is heavily calcified, especially along the proximal border. There is no exterior sign of attachment on most of them. Instead, the heavy calcification in the proximal region engulfs the base and all nearby branches.

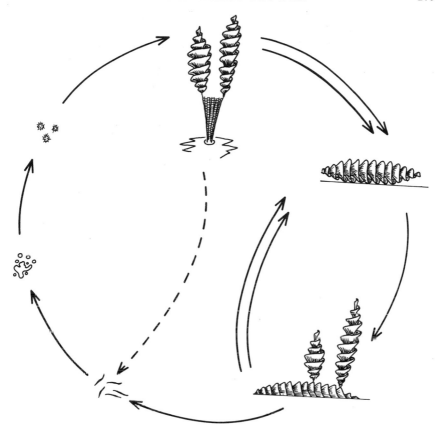

Figure 8.10 Life history of *Archimedes*, beginning with gametes at lower left, proceeding clockwise through inferred polyembryonic fission of embryos, free-swimming larvae, to colonies at top and right. Repeated breakage of colony bases and subsequent growth of secondary spirals from distal portions of toppled colonies could lead to enormous local clonal populations. (After McKinney 1983, courtesy of the Paleontological Society.)

The form of larger colonies developed after two distinct phases of growth. While still erect at an early growth stage, colonies developed a transverse bow with zooidal apertures on the convex side, and apparently grew so that zooidal apertures faced the current. Then, when the current eventually pushed them over, the colonies landed in such a way that apertures faced away from the substratum and the bow of the fan caused the colony to rest on its lateral margins, with space between the substratum and the reverse side of the fan for outflow of filtered water. For those colonies that fell in the most favorable position, the growing edge of the fan extended approximately in the direction of flow (Fig. 6.20), while the upstream proximal edge was progressively more weighted by skeletal deposition. The colonies generally developed paraboloid shapes that presumably offered minimal interference to the flow of water and were stabilized by their weight distribution. The growing edge appears to have tracked the boundary layer between actively flowing water above and the current shadow on the downstream end. This would have allowed zooids to feed from rapidly moving water while situated in quieter conditions, much as sheets of *Membranipora* on exposed surfaces of *Laminaria* do today (Lidgard 1981). Presence of lophophores on the convex surface of *Lyropora* and *Lyroporella* contrasts with the position of feeding members of fan-shaped colonies that are passive suspension feeders and are adapted to unidirectional flow. For example, shallow, concave, erect fans of the flexible octocoral *Melithaea* and of the hydroid *Aglaophenia* have polyps extending only from the concave side (Wainwright *et al.* 1976) and feed from the turbulent, more slowly moving water on the concave downcurrent side in unidirectional flow.

Cilia-generated feeding currents of bryozoans that venture out of the boundary layer should interact with any appreciable ambient flow, since they flow at only a few mm sec^{-1} (Lidgard 1981, Best & Thorpe 1983, 1986a, b). This has been demonstrated for two species of *Bugula*. Increased penetration of high flow velocities with increased ambient velocity displaces the feeding success of zooids from upcurrent to downcurrent portions of *B. stolonifera* colonies (Okamura 1985). Zooids of *B. neritina* feed on upcurrent, downcurrent, and lateral sides of colonies at ambient velocities up to about 4 cm sec^{-1} (although the bell shape of lophophores is obviously distorted by the through-flowing water), and progressively higher ambient velocities inhibit the cilia-generated flow of an increasing number of zooids in exposed positions (McKinney *et al.* 1986a). At 15 cm sec^{-1} ambient flow, zooids in 3–4 cm diameter colonies appear helpless and do not open on upcurrent and lateral sides, and those on downcurrent sides stream out in much narrower bells than normal, with no visible catching or trapping of food particles that race by, and little catching of particles from the turbulent zone in the leeward shadow. Colonies of *B. stolonifera* and *B. neritina* live in shallow, vigorously flowing and waveswept water and probably catch most of their food during the travel between maximum "to" and "fro" positions, during which the zooids are travelling with the water and therefore experience lower flow velocity than the ambient current, as suggested for flexible gorgonians such as *Pseudopterogorgia americana* (Wainwright *et al.* 1976).

8.3.2   Rigidly erect colonies: adeoniform

Erect, branched, bilaminate colonies are unique in having symmetrically flattened or ovoid branch cross sections, branch widths of about 1–5 mm depending upon species, narrowly varying branch bifurcation angles within a colony, set lengths of branch links between successive bifurcations in most, and generally some degree of twisting that accompanies bifurcation (§ 3.2.2). These basic morphological attributes result in strongly asymmetrical mechanical response when colonies are small, but in many species growth reduces the degree of anisometric behavior with increased colony size.

The mechanical response of bilaminates to current-generated drag and to impact with other solid objects has been extensively studied for the adeoniform cheilostomes by Cheetham and colleagues (Cheetham *et al.* 1981, Cheetham & Thomsen 1981, Cheetham & Hayek 1983, Cheetham 1986a). Of particular interest are several trends in mechanical properties through time that appear to reflect improved adaptation to the physical environment.

In most adeoniform species, branches in the proximal portions of colonies thicken while branch tips lengthen and divide. This increase in thickness occurs by thickening of frontal shields, by deposition of extrazooidal skeleton across the surfaces of zooids, or by frontal budding of additional layers of zooids. Among nine living adeoniform species investigated, all but one produce such tapering branches and have zooidal surfaces occluded by extrazooidal skeleton in proximal parts of colonies. Average proximal thickening has increased steadily in adeoniform cheilostomes since the Cretaceous (Table 8.1, Fig. 8.11).

Cheetham and Thomsen (1981) found no significant differences in modulus of rupture (*Mr*) between various combinations of eight modern adeoniform species arranged by microstructure and mineral composition (i.e. calcite vs. aragonite), but did find significant differences ($P < 0.001$) between species within such groups, regardless of how arranged. Thus, variations in skeletal materials do not seem to be important determinants of colony strength of adeoniform colonies.

Strength expressed as *Mr* is relatively high in modern adeoniform cheilostomes (dry *Mr* varies from 22.5–70.6 meganewtons m^{-2}; $\bar{X} = 42.1$ meganewtons m^{-2}; corrected values of Cheetham & Thomsen 1981), and is closer to that of well-calcified mollusks than to that of cyclostomes and scleractinians, which have virtually all-mineral skeletons. This would seem to give adeoniform cheilostomes greater durability as erect benthic colonies compared to cyclostomes and scleractinians of similar branch and colony

Table 8.1 Averages for values of branch thickness for Recent, Oligocene, and Paleocene species of adeoniform cheilostomes that exhibit proximal thickening (summarized from Cheetham *et al.* 1981). Thickness at growing tip (t_{gt}) is the Y-intercept of regression lines; coefficient *b* represents the thickening gradient in the distal part of the colony and is the slope of the regression line. Thickening is significantly greater ($P < 0.001$) in Recent species than in the fossil species.

Age	Number of species	Number of specimens	*b*	t_{gt} (mm)
Recent	8	203	0.036	0.676
Oligocene	5	444	0.021	0.662
Paleocene	4	695	0.020	0.546

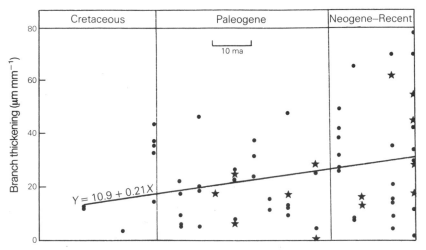

Figure 8.11 Increase in gradient of branch thickening in adeoniform cheilostomes from Cretaceous to Recent. (After Cheetham 1986a, courtesy of the Royal Society, London.)

size. However, apparently there has not been an increase in material strength of adeoniform cheilostomes through time. To the contrary, Cheetham and Thomsen concluded that *Mr* of adeoniform skeletons has *decreased* through the Cenozoic. For this analysis, *Mr* of modern species was determined for colonies in wet condition, dry condition, and with the organic component removed. *Mr* of fossil species was determined for colonies varyingly affected by diagenesis, and corrected for decreases in *Mr* observed in all comparisons between fossil and modern specimens within a species or genus. In addition to the overall decline in *Mr* through the Cenozoic, the presumed most primitive living species examined, *Membranipora savartii*, has the highest *Mr* among modern adeoniform species.

Despite the decrease in material strength of adeoniform cheilostomes through the Cenozoic, resistance of colonies to breakage increased over the same period. This is because resistance to breakage under various types and magnitudes of forces depends upon both bending strength and factors of morphologic design. Cheetham and Thomsen found that there is an inverse relationship between bending strength and design in adeoniform cheilostomes: the most resistant designs are found in species with low *Mr*, and the least resistant are in species with high *Mr*.

Proximal thickening by zooidal or extrazooidal skeleton constitutes the principal improvement. The initial flattened morphology at branch tips results in greater resistance to breaking when stress is applied from an edge than when applied to one of the faces of a branch. However, branch thickening, where it occurs, is much greater perpendicular to the median plane. In basal parts of large colonies, branches may reach radial symmetry both in morphology and in bending moment to rupture. Radial symmetry in resistance to breakage is beneficial near bases of large colonies because the colonies themselves approach radial symmetry with increase in size, and therefore are likely to receive substantial stress from any direction as they grow higher into more vigorously moving water.

Branch thickening presumably costs energy for skeletal secretion and for transportation of nutrients from feeding parts of the colony, in addition to loss by self-overgrowth of proximal zooids which might otherwise feed and reproduce. Exponential increase in surface area with increased branch length helps to offset the magnitude of this loss; Cheetham & Hayek (1983) noted that even where orifices are occluded in the inner 60% of branch length in *Cystisella saccata*, only 20% of the total zooids in the colony are lost. Loss of feeding and reproduction by structural modification of proximal parts of erect branched colonies is not restricted to adeoniform species, but is widespread in other erect bryozoans, in gorgonian corals, and in stylasterine corals.

There is a consistent buildup of stress generated by drag forces, such as would be generated by a uniformly flowing current, towards the colony base in all adeoniform species. This is a result of the geometric increase in cross-sectional area of the colony with growth, and the progressive concentration downward through branch junctions of stress from all across the colony. This is partially ameliorated in colonies whose branches thicken proximally, but it increases greatly in colonies that do not thicken basally (Fig. 8.12, curve set A). For example, stress at the base of the largest *Membranipora savartii* colonies is up to 100 times the stress at the earliest stage of erect growth.

In species in which branches thicken with age, stress due to even distribution of drag force across the whole colony should be greatest just above the colony base (Fig. 8.12, curve sets B and C). Thus when current velocity and drag force increase sufficiently, the colony will tend to break off near the base and all, or all but a stump of it, will roll away, frequently colliding with other objects which may cause concentrated loading of force at or near branch tips.

Force applied to individual branch tips of attached colonies causes a bending moment that also increases toward the colony base, but the magnitude of such local loading of force depends upon the size and velocity of a colliding object or the force applied by a bite, and not upon the colony area. Thus, proximal thickening of branches should have a better chance of reducing stress and breakage near the colony base due to concentrated loads than that due to uniform loads. Most species of adeoniform cheilostomes have a distinct region where stress reaches a maximum due to impact, which is either at the colony base or more commonly near the branch tips (Fig. 8.13). The latter, more common design (Fig. 8.13, curves B and C), seems to ensure that damage to a colony caused by local impact is minimal, resulting only in loss of branch tips.

Young colonies of most species can withstand exceedingly high currents (greater than $3\,\mathrm{m\,sec^{-1}}$), but minimum velocity that would cause stress

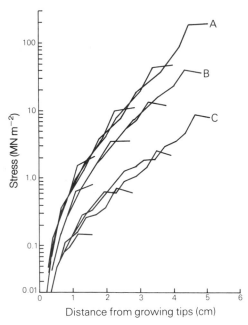

Figure 8.12 Distribution of stresses in adeoniform colonies due to a uniform load of $20 \, mg \, mm^{-2}$ across entire colonies, as calculated by Cheetham & Thomsen (1981). Curve set A represents species, such as *Stomachetosella crassicollis*, in which the proximal thickening gradient is low; curve set B represents species, such as *Kleidionella verrucosa*, in which proximal thickening gradients are intermediate; and curve set C represents species, such as *Cystisella saccata*, in which proximal thickening gradients are high. (From Cheetham & Thomsen 1981, courtesy of the Paleontological Society.)

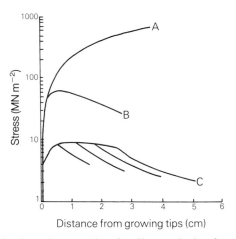

Figure 8.13 Distribution of stresses in adeoniform colonies due to a static load of $50 \, g$ concentrated at one growing tip, as calculated by Cheetham & Thomsen (1981). Curve A represents species, such as *Euritina torta*, in which proximal thickening gradients are low; curve B represents species, such as *Bracebridgia subsulcata*, in which proximal thickening gradients are intermediate; and curve set C represents species, such as *Cystisella saccata*, in which proximal thickening gradients are high. (From Cheetham & Thomsen 1981, courtesy of the Paleontological Society.)

sufficient for breakage diminishes rapidly with increasing colony size. In most modern species, breakage due to drag force is projected to occur only at velocities above $1 \, m \, sec^{-1}$, even for the largest colonies. This contrasts with Oligocene and Paleocene species which, on average, appear to have broken at lower velocities than those that damage modern species, regardless of colony size.

Cheetham and Thomsen also calculated the additional load at branch tips sufficient to break colonies when applied in the same direction as a uniform load (e.g. drag). Under natural conditions this could be provided by a shell fragment, beer bottle, or other object bouncing along the sea floor with the current. Among the 18 species of fossil and Recent adeoniform species considered, there has been a striking increase through the Cenozoic in the average additional load needed to break colonies at any velocity, and resistance to breakage at any velocity for the strongest of the species (Fig. 8.14). During the Paleocene no colonies could resist breaking at a concentrated load of 50 g at branch tips, but one Oligocene and two modern species did not break even at concentrated loads of over 100 g at branch tips.

Two contrasting modes of design seem to have evolved among adeoniform taxa. The most common is an apparently increased structural integrity in which colonies resist breakage and remain attached to their original point of growth. This is suggested by the overall Paleocene to Recent increase in resistance of colonies to breakage due to changes in branch thickness, link length, and branch width. For example, the species most resistant to breakage among those studied so far is *Cystisella saccata*, which is common on the Grand Banks, where $1-2 \, m \, sec^{-1}$ bottom currents may be generated to depths of 50–100 m during storms. At those depths, *C. saccata* colonies may exceed 5 cm height, approaching their probable maximum attainable size. Judging from museum specimens (there are no field studies), *C. saccata* colonies on the Grand Banks generally remain intact and attached to their original substratum, even though they are built of the weakest skeletal material so far found among adeoniform cheilostomes.

The alternate design is to decrease structural integrity so that breakage and regrowth from fragments are common. For example, Paleocene *Coscinopleura digitata*, a species with a structural design poorly resistant to breakage, commonly regenerated from broken fragments (Cheetham *et al.* 1981). As in *Cystisella saccata*, impact-generated stress maxima are near branch tips. However, although modulus of rupture is about equal for the two species, *Conscinopleura digitata* is much less robust than *C. saccata*. *C. digitata* is "superabundant" in the Vincentown Formation, apparently because breakage produced viable fragments that continued to grow and establish new colonies. Fragmentation has led to fantastic abundances of many other erect bryozoans including Paleozoic rhabdomesid cryptostomes (Blake 1976), trepostomes (Ettensohn *et al.* 1986), the fenestrate *Archimedes* (McKinney 1983), and living cellarinellid cheilostomes (Winston 1983).

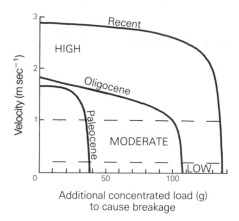

Figure 8.14 Calculated curves for concentrated loads that, at various flow velocities, would break Stage 3 colonies of the most resistant adeoniform species examined from the Paleocene, Oligocene, and Recent. (After Cheetham & Thomsen 1981, courtesy of the Paleontological Society.)

8.3.3   Rigidly erect colonies: radial

Erect colonies with radially symmetrical branches can give rise to new branches in any plane that passes through the axis of the parent branch, all-in-all a more flexible arrangement than the fixed planes in which unilaminate and adeoniform branches can divide. Colonies with radial branches have the potential to develop three-dimensional bushy shapes as soon as they begin to extend away from the substratum. In addition, radial branches may be of any diameter from about 0.2 mm up to at least 40 mm, and their diameters in some colonies and species are highly variable. A possible disadvantage of radial branches is the greater difficulty, or necessity for more elaborate structures, in elimination of filtered water than characterizes unilaminate and adeoniform colonies. Also, less highly regulated branching may result in more frequent interference, fusion, or termination of branches than occurs in the other branching types.

There have been no studies on strength and design of bryozoans with radial branches comparable to those on bilaminate colonies. A field study of the environment of arborescent *Heteropora pacifica* with radial branches (Fig. 4.16) demonstrated currents of up to 125 cm sec^{-1} flowing 6–10 cm above the substratum (Schopf *et al.* 1980). However, even the most robust *H. pacifica* colonies studied probably could not have survived stress generated by such flow (unpublished calculations). They existed in the environments examined because they were less than 2 cm high, thereby living within the attenuated flow characteristic of boundary layers.

Biomechanical studies of branched corals by Chamberlain and Graus (1975), however, bear on understanding the flow regime of arborescent radially branched bryozoans. Both have roughly similar colony forms and generally similar dry strength (Chamberlain 1978, Schopf *et al.* 1980). Water approaching the upper and lateral portions of a branched coral flows over and around the colony without entering the interior, whereas that water approaching the midpoint of a colony's profile as presented to the current, enters the interspaces between branches; but nearly all exits from the colony at the top rather than at the back (Fig. 8.15). As the water moves through the colony, it is dispersed and mixed by contact with interior branches, so that water leaving the colony is more turbulent than when it entered.

Both external flow velocity and growth form affect the rate and path of flow through branched coral colonies. Change in external flow velocity causes changes in acceleration of water around the colony exterior, interior deceleration and mixing, and dispersion of the water leaving the colony. Water entering a colony at relatively high velocity can penetrate farther into the interior than at relatively lower velocity (Fig. 8.15), but in no case does the region of relative stagnation within the interior entirely disappear.

Growth form characteristics that influence flow include colony porosity, relative branch size, branching pattern, and branch alignment. Water movement through a colony depends in large part on porosity or openness of the framework, which is determined by spacing of nearest neighbor

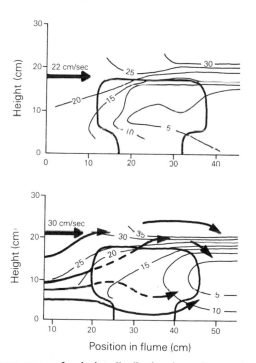

Figure 8.15 Contour maps of velocity distribution in and around a specimen of the scleractinian coral *Oculina diffusa* at two ambient flow velocities. Heavy arrows in the lower diagram plot flow directions through the specimen. (After Chamberlain & Graus 1975, courtesy of the University of Miami Rosensticl School of Marine Science.)

branches and by average branch diameter. Colonies with open framework and high porosity, such as the scleractinian *Acropora cervicornis*, do not readily develop a pronounced diagonally-outward flow pattern, because loss of momentum occurs slowly for incoming water, which can therefore penetrate far inside before decelerating or being forced out.

Relative branch size determines the texture of the profile that an erect branched colony presents to a current. With similar branch arrangements and colony porosity, colonies with widely spaced but larger branches may offer relatively less interaction with the flow, thereby generating less drag, than colonies with more closely spaced but thinner branches. Water can therefore enter and travel farther in colonies with large branches than in those with similar skeletal volume but thinner branches.

In addition, regularity of placement and alignment of branches within a colony determines nature of flow through it. A model consisting of randomly placed vertical dowels, with irregularly distributed clumps and holes, clears laterally flowing water very erratically, with some rapidly flowing streams in channels, and stagnant areas behind clusters of closely spaced dowels. In contrast, a model with equivalent porosity and branch size, but with evenly spaced branches, allows rapid ventilation throughout, even within the relatively stagnant downstream region (Chamberlain & Graus 1975).

The various attributes of colony form can vary so that dynamic similitude exists for very differently constructed colonies. This variation may occur in complex ways with changes in all attributes, but it can be most easily grasped where variations in two parameters offset one another so that no change in flow state results. For example, flow within the corals *Oculina diffusa* and *Stylophora* sp. is strikingly similar, despite considerable differences in form (Fig. 8.16). Approximate dynamic similitude exists between them: the differences in porosity and in branch diameter approximately offset one another. This suggests that more than one morphological solution to a specific hydrodynamic problem may exist, and that a considerable range in colony form may be suited for any specific hydrologic situation, provided that the structures are capable of absorbing the stresses generated or of coping with damage.

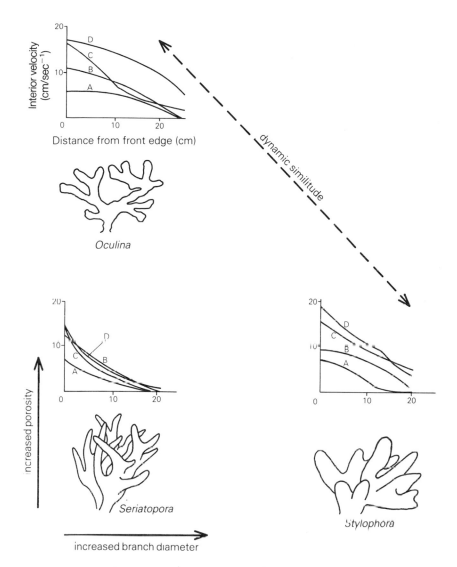

Figure 8.16 Dynamic similitude in flow through colonies of the corals *Oculina*, which has relatively small branches and high porosity, and *Stylophora*, which has relatively large branches and low porosity, contrasted with that of *Seriatopora*, which has relatively small branches and low porosity and is relatively more impermeable than the other two. Curves A–D represent successively higher ambient flow velocities. Axes of each graph plot the same variables. (After Chamberlain & Graus 1975, courtesy of the University of Miami Rosenstiel School of Marine Science.)

9 Life on and in sediments: problems of substratum stability

From an ecological perspective, solid substrata are surfaces suitable for attachment and growth of organisms, the size of which is typically larger than their inhabitants. Sediments, in contrast, are masses of particles of similar or smaller size than their inhabitants. Because they are smaller, sedimentary particles are also more easily moved about than hard substrata. Thus, life on sediments poses problems of colony stability and support not usually faced by sessile species on hard substrata.

There are three general life habits of bryozoans on sediments: free-living, rooted, and interstitial. In each case these include species that are specifically adapted to life on sediments as well as numerous forms more characteristic of solid substrata. All of the specialized groups are polyphyletic and show distinctive convergence in form and correlation with specific environments even more remarkable than among sessile species. They also share a strong tendency to assume traits characteristic of solitary animals, including compact morphology and determinate growth. Moreover, like solitary animals in unstable habitats, many have precocious sexual reproduction. This trend towards a solitary strategy in byrozoans parallels the general dominance of solitary over colonial animals on sediments (Jackson 1985).

More than any other bryozoans, sediment-dwelling species also share the unfortunate characteristic of a firmly rooted descriptive jargon based on higher taxonomic names. This causes confusion because the terms are unnecessarily complex, and many members of the same taxon commonly grow in different forms (e.g., "conescharelliniform" Orbituliporidae and "orbituliporiform" Conescharellinidae!). More importantly, these terms often obscure interesting parallels in life habits, as between free-living "lunulitiform" and "setoseliniform" species. We have generally avoided this terminology except where necessary to avoid confusion with the established literature.

9.1 Free-living bryozoans

Erect bryozoans commonly may be detached from their substrata by physical processes or predators. If they are large enough to be stable lying free on the bottom, or are able to reattach, such fragments may survive to build up dense populations of asexually produced colonies (§ 8.3.1, Winston 1983). Bryozoans may also settle on minute substrata such as sand grains, foraminiferans, or shell fragments which they may eventually outgrow (§ 7.1). If encrusting, they may form increasingly massive, roughly spherical, passively mobile colonies which enlarge by self-overgrowth or frontal budding (Fig. 9.1, Rider & Enrico 1979, Balson & Taylor 1982). Similarly, erect bryozoans that outgrow their initial substrata may simply topple over and form, with their own dead skeleton, a larger, more stable substratum for their continued upwards growth. Both these forms of growth are a consequence of, rather than a specific adaptation to, life on sediments, and have been common among stenolaemate and gymnolaemate bryozoans throughout their history.

Large, sheet-like colonies were common on the surfaces of sediments from the Middle Ordovician until the Carboniferous. They typically began growth on brachiopod shells or other skeletal debris and spread across the surrounding sediments, covering an area up to several times the radius of the original substratum. This growth form was common among trepostomes

1 cm

(a)

(b)

(c)

(d)

Figure 9.1 Circumrotatory growth through the Phanerozoic: (a) *Conopeum* sp., Recent, North Carolina; (b) *Stylopoma spongites*, Waccamaw Formation (Pleistocene), North Carolina; (c) intergrowth *Porella*, *Conopeum*, *Vibracellina*, and others, Castle Hayne Formation (Eocene), North Carolina; (d) *Fistulipora* sp., Lower Silurian, Alabama.

(e.g., *Nicholsonella, Tabulipora*) and cystoporates (*Ceramopora, Fistulipora*) and was best developed in Silurian *Lichenalia*, which formed disc-shaped colonies up to at least 50 mm radius and less than 1 mm thick (Hall 1852, Bassler 1906). Continued thickening of colonies, either by elongation of zooids or self-overgrowth, and accompanied by lateral growth, resulted in hemispherical free-lying colonies characteristic of several trepostomes (e.g., *Prasopora, Mesotrypa*). Not much more is known about these free-lying colonies because there has been no detailed study of the phenomenon, and systematic descriptions are typically vague regarding the exact shape of colonies, extent of attachment, and type of substratum and sediments.

The demise of passive free-lying bryozoans in the late Paleozoic (Fig. 4.18) coincides with increased bioturbation since that time (Thayer 1979, 1983). From the Late Carboniferous to Jurassic, the number of taxa that include bioturbators increased from 13 to 32, and bioturbation is estimated to have increased by roughly an order of magnitude. Re-establishment of an abundant and diverse bryozoan fauna during the Mesozoic did not include passive free-lying forms.

Instead, assemblages of free-living bryozoans have been dominated ever since the Upper Cretaceous by small discoidal, cup-shaped, or conical colonies which lack rootlets for attachments and live free on or just below the sediment surface (Figs. 9.2, 9.3). All are cheilostomes, and the great majority are anascans. These species, commonly termed **lunulitiform**, exhibit a unique set of morphological and behavioral adaptations for life on sediments. There are two major groups, the Lunulitidae and Cupuladriidae (Cook & Chimonides 1983). The former became abundant in Europe and Asia during the Upper Cretaceous, and in North America in the Paleocene. They disappeared from the Americas by the end of the Miocene and from Europe by the end of the Pliocene, and are now restricted to Australasia, where they are still abundant. The Cupuladriidae first appeared in the Paleocene of western Africa whence they spread to Europe and Asia by the Oligocene and the Americas by the Miocene. They are abundant today in suitable habitats in the tropical and subtropical Atlantic, Pacific, and Indian Oceans, but are found in only the extreme south-western end of the Mediterranean (Lagaaij 1963).

The habits and behavior of the two families are so convergent that the following account suffices for both. Lunulitids occur on sediments ranging from coarse sands to muddy bottoms but cannot tolerate high turbulence or extensive deposition of silt (Marcus & Marcus 1962, Rucker 1967, Cadee 1975, Cook & Chimonides 1983). They are found as shallow as 2 m on sandy shoals swept by waves or tidal currents of about 100 cm sec^{-1}, to depths greater than 500 m, but are most abundant on continental shelves between about 20 to 70 m where population densities are reported to reach 15,000 per m^2. They have been collected alive only from warm waters (10–29°C) between 14–37‰ salinity.

9.1.1 Morphology and behavior

Colonies of both families of lunulitids are supported and stabilized upon the sediment by the long mandibles of peripheral and subperipheral avicularia (Fig. 9.3, Marcus & Marcus 1962, Cook 1963, Greeley 1967, Cook & Chimonides 1978). The normal life position is with the convex frontal surface upwards. Setae on the frontal surface of colonies sweep away sediment particles, even to the extent of enabling colonies to reach the surface of the sediment following sudden burial. Despite their supportive role, there is no relation between their length and the diameters of colonies

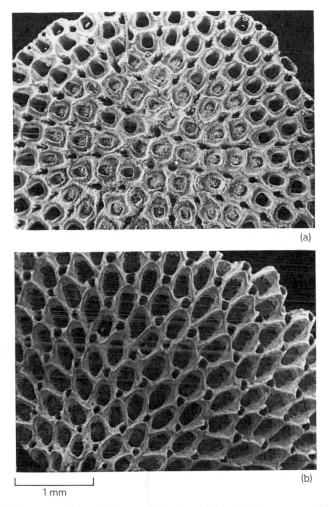

(a)

(b)

1 mm

Figure 9.2 Representatives of the two families of lunulitid bryozoans: (a) *Lunulites jacksonensis*, Lunulitidae; (b) *Cupuladria canariensis*, Cupuladriidae. (Photographs courtesy P. L. Cook and P. J. Chimonides. Reproduced by permission of the British Museum (Natural History).)

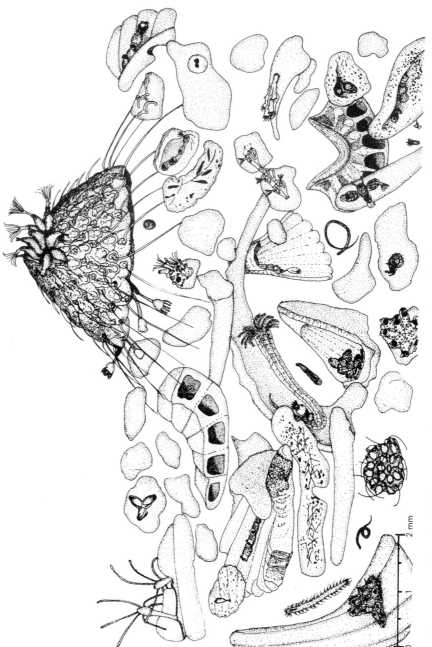

Figure 9.3 Diagrammatic cross-sectional view of the sand fauna of Capron Shoals, Florida. At the surface is *Cupuladria doma*, supported by long peripheral mandibles. The colony is "fouled" by a ctenostome and entoprocts. Sand grains are encrusted by bryozoans, worms, and hydroids. (Courtesy of J. E. Winston.)

in either group. There are, however, considerable differences in the design and function of avicularia (Cook 1963, Cook & Chimonides 1978). Avicularia of Lunulitidae are budded as interzooids scattered over the frontal surface or arranged in radial rows (Fig. 9.2). Primitively, the articulation points that support the mandible are simple and symmetrical so that the mandible can swing only in two directions in a single plane. Many Recent species, however, have more complex, tilted articulation of the mandibles that allows for movement over a wide degree of arc (Fig. 9.4). Mandibles are varyingly smooth or serrated, bristle-like or paddle-shaped. In contrast, avicularia of Cupuladriidae develop from a common

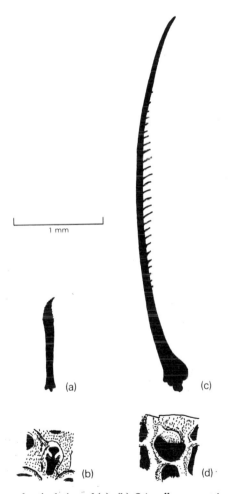

Figure 9.4 Mandibles and articulation of (a), (b) *Otionella symmetrica* and (c), (d) *Selenaria maculata* ((a) from Cook & Chimonides 1984a, courtesy of the British Museum (Natural History); (c) drawn after Cook & Chimonides 1978. Reproduced by permission of the British Museum (Natural History).)

autozooecial–avicularian bud. Mandibles are oriented alternately asymmet-
rically to right or left within each radial row of zooids (Fig. 9.2). They are
long, bristle-like, sigmoid, and can only swing in two directions of a single
plane (Fig. 9.5). Nevertheless, because of their curvature and asymmetrical
articulation, they collectively are able to sweep over most of the frontal
surface of the colony.

The mobility of four Cupuladriidae and two Lunulitidae has been studied
in the laboratory (Marcus & Marcus 1962, Cook 1963, Greeley 1967, Cook
& Chimonides 1978). The behavior of the cupuladriids is similar. Colonies
can maintain themselves on the sediment, regain the surface after burial,
and reposition themselves if turned over, all by movements of their
mandibles. Colonies are also frequently turned over by other animals in
aquaria, particularly hermit crabs that feed on cupuladriid-borne epiphytic
algae, and the same almost certainly occurs in nature.

The behavior of the two Lunulitidae is much more varied. *Otionella
symmetrica* grows to 7 mm diameter and has short (0.7 mm), spatulate

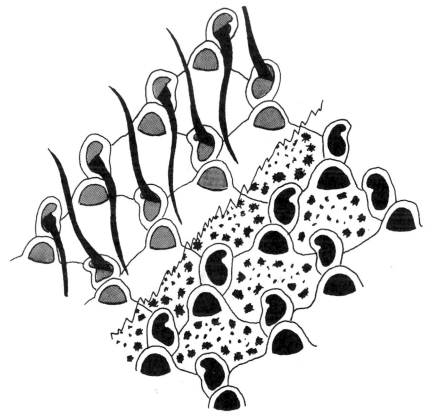

Figure 9.5 Orientation and articulation of mandibles of *Cupuladria umbellata*. Upper left
with cuticle and mandibular setae in place, lower right shows skeleton with cuticle and setae
removed. (Drawn after Cook 1965a.)

peripheral mandibles (Fig. 9.4a) that maintain the colony about 0.5 mm above the substratum. Movement of the mandibles is restricted to a single plane (Fig. 9.4b). This species can exhume itself but cannot turn over unassisted. *Selenaria maculata* is larger (13 mm diameter) and has peripheral mandibles 4 mm long. These are setiform, flexible, and serrated on one side (Fig. 9.4c); they have complex, asymmetrical articulation which allows movement in nearly all planes and directions (Fig. 9.4d). Movement of mandibles is more rapid than in other species, and is often highly coordinated. This not only allows colonies to turn themselves over (Fig. 9.6) but to travel over the surface of the substratum by a series of lurching movements at speeds up to $1\,\mathrm{m\,h^{-1}}$. Such movements are particularly stimulated by blue-green light (Chimonides & Cook 1981).

The differences in behavior of the two Lunulitidae are very clearly reflected in their skeletons. On the other hand, the four cupuladriids are quite distinct morphologically from *Otionella symmetrica*, and yet their behaviors are rather similar. Observations of many more species, comparable to those of feeding behavior (§ 6.3), are necessary to evaluate the potential for paleobehavioral studies of fossil free-living species.

Species of Lunulitidae and Cupuladriidae are commonly fouled on their frontal surfaces by algae and on their basal surfaces by a variety of

Figure 9.6 Acrobatics of *Selenaria maculata*: (a) living colony viewed from above; (b) lateral view of colony moving over a glass surface; (c) two colonies on sand, colony to the left starting to climb over the one to the right; (d) overturned colony with peripheral setae pointed downward and epibionts on basal surface; (e) colony within 2 sec of reattaining its "normal" position. (From Cook & Chimonides 1978, courtesy of the British Museum (Natural History) and *Cahiers de Biologie Marine*.)

invertebrates including tube worms, bivalves, entoprocts, other bryozoans, and sponges (Figs 9.3, 9.6d, 9.7, Cook 1963, 1965b, Greeley 1967, Chimonides & Cook 1981). Many animals attached to basal surfaces seem to have little effect on their hosts except to weigh them down. Possible exceptions are sponges that may disrupt normal budding patterns, excavating ctenostomes, and hydroids whose stolons grow up onto the frontal surfaces of colonies (Cook 1965b). On the other hand, tubiculous isopods living on the basal surface of *Selenaria maculata* have been observed in the laboratory to feed on algae growing on their hosts, and may be important cleaning symbionts in nature (Chimonides & Cook 1981). Hermit crabs also clean algae from free-living bryozoans (Greeley 1967, Cook 1986); the crabs pick off the algae but may also strip off the basal membrane or even break the colonies they graze.

The idea that epibionts may be detrimental to free-living bryozoans is strongly supported by the discovery of molting by *Cupuladria doma* (Winston & Håkansson, in preparation). In shallow water, this species is often thickly overgrown by algae, along with the usual assortment of animals (Fig. 9.7). Under these conditions, and perhaps at other times as well, the frontal membrane cracks off revealing a clean, new membrane beneath (Fig. 9.8). This is not a growth process as in crustaceans and therefore seems to be an adaptation for cleaning. It is not known whether molting occurs in other bryozoans.

9.1.2   Demography and life history

Like solitary animals, free-living bryozoans are easily counted and measured, and are thus readily amenable to demographic study (Håkansson 1975). The most detailed study of Recent species is for *Cupuladria doma* and *Discoporella umbellata depressa* from about 6 m depth on a sandy shoal near Fort Pierce, Florida (Håkansson & Winston 1985, Winston 1986). Apparent densities (juveniles excluded) on the sand surface are low: 0.5 per m² for the former species and 0.1 per m² for the latter. Size-frequency distributions of *C. doma* peak at about 2.5 mm diameter, with a maximum of about 5 mm (Fig. 9.9). At this upper limit budding of new autozooids stops, and a terminal double row of vibraculae is formed. Most colonies suffer breakage along their terminal margins and die during winter storms; thus *C. doma* may effectively be an annual species at this site. Size-frequency distributions of *D. u. depressa* are more variable, with peaks ranging between 1.0 and 2.5 mm. The upper limit is about 7 mm (Fig. 9.9), which is only one-tenth the maximum of 70 mm reported for this species (Cook 1986). Colonies apparently live more than one year.

Colonies of *C. doma* are small but robust, with extensive basal thickening. Damage to edges is common, but breakage of entire colonies is rare, and only 4–8% of colonies show signs of major regeneration. Colonies of *D. u. depressa* are less robust, and asexually produced colonies derived from fragmentation and regeneration make up 60% of all the colonies in the population. Asexual production of colonies is also common in Brazilian populations of this species, where it occurs both by fragmentation and a process termed zoarial budding (Fig. 5.18, Marcus & Marcus 1962). Budding was observed in populations collected from 4–32 m depth, but not below. Colonies producing buds ranged between 4 and 9 mm in diameter. As many as 15 buds were attached to a single colony. Zoarial budding commences with the formation of a pseudoancestrular zooid from which the

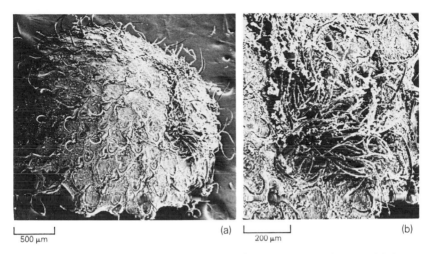

(a)   (b)

500 μm   200 μm

Figure 9.7 Epibionts on *Cupuladria doma*: (a) view of upper surface of colony, (b) close-up of tuft of filamentous algae overgrowing several zooids. (Courtesy of J. E. Winston.)

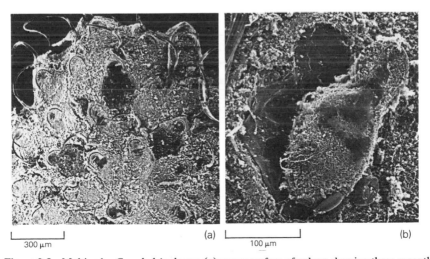

(a)   (b)

300 μm   100 μm

Figure 9.8 Molting by *Cupuladria doma*: (a) upper surface of colony showing three recently molted zooids, (b) close-up of single zooid showing partially detached operculum and frontal membrane. (Photographs courtesy of J. E. Winston.)

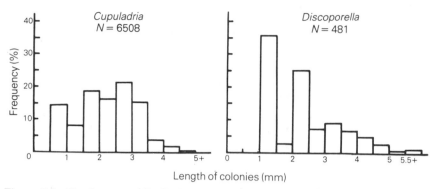

Length of colonies (mm)

Figure 9.9 Size-frequency distribution of *Cupuladria doma* and *Discoporella umbellata depressa* at Capron Shoals, Florida, summed over one year. (From Winston 1986, courtesy of the National Geographic Society.)

rest of the daughter colony is derived. These commonly grew to 10–30 zooids before detachment by breakage of the pseudoancestrula. In populations off western Africa, 80% of colonies of the closely related *D. umbellata* were derived from fragments, as were all of the approximately 100 *O. symmetrica* colonies studied by Cook & Chimonides (1978). Many Recent and fossil populations are derived entirely from fragments (Hakansson & Thomsen 1985). Fragments as small as 2–3 zooids may successfully regenerate new colonies (Cook & Chimonides 1983, 1984a).

Most species of Lunulitidae stop growing at some approximately determinate size limit. The margin is then finished off by budding of varyingly distinctive peripheral zooids, much as in *C. doma* discussed above. In addition, peripheral autozooids of many Lunulitidae (e.g., *Otionella*, several *Selenaria*) are larger, have larger opesia, or are differently shaped than their older neighbors (Fig. 9.10, Chimonides & Cook 1981, Cook & Chimonides 1985a, b). These peripheral polymorphs are involved in sexual reproduction. In *S. maculata* and related species, subperipheral female zooids are bounded by peripheral male zooids and enlarged avicularia (Figs. 9.10, 9.11). Embryos are brooded within ovisacs within the enlarged female zooids. Male zooids possess modified, nonfeeding lophophores with only two tentacles; these are 2 mm long and extend from tentacle sheaths an additional 4 mm. Sperm are released from the tips of the tentacles a full 6 mm from the edge of the colony. In relation to this zonation of sexual function, feeding in mature *S. maculata* is limited to a ring of zooids within the sexual peripheral region; the central zooids occupying about one-third of the surface have no lophophores. This pattern is opposite that characteristic of encrusting species in which peripheral zooids are most susceptible to overgrowth by other organisms. Larval behavior has not been observed. A single ancestrular zooid is formed at metamorphosis (Fig. 9.12c) which usually occurs on a sand grain, foraminiferan, or small fragment of shell. Some fossil species may have settled without any initial attachment

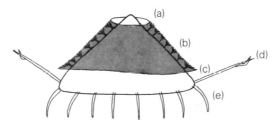

Figure 9.10 Distribution of heterozooids in *Selenaria bimorphocella*. Peripheral male zooids (m) and avicularium (a), subperipheral female zooids (f) and autozooids (z). (From Chimonides & Cook 1981, courtesy of the British Museum of Natural History) and *Cahiers de Biologie Marine*.)

Figure 9.11 Side view of *Selenaria maculata* showing zonation of functions: (a) nonfeeding zooids around ancestrula; (b) feeding autozooids; (c) brooding zooids, may feed or not; (d) nonfeeding male zooids with elongated tentacle sheaths for release of sperm; (e) enlarged peripheral avicularia. (Drawn after Chimonides & Cook 1981.)

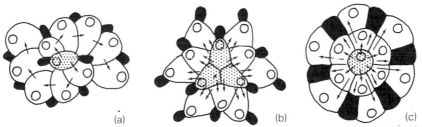

Figure 9.12 Early astogeny of lunulitiform bryozoans: (a) *Helixotionella*, Lunulitidae; (b) *Cupuladria*, Cupuladriidae; (c) *Lunulites*, Lunulitidae. (Drawn after Cook 1965a.)

(Håkansson 1975) as has been shown for living *Lunularia* (Cook & Chimonides 1986).

Growth of cupuladriids is also approximately determinate, but sexually polymorphic zooids and ovicells are absent (Cook 1965b). Embryos are brooded within autozooids. Larval release by *C. doma* has been observed throughout the year in aquaria (Winston 1986). Larvae settled quickly near their parents; some crawled over the substratum for up to two hours, but many metamorphosed under, or adjacent to, adult colonies, possibly their parents. A triadic, multizooidal ancestrular complex is formed at metamorphosis (Fig. 9.12b), which usually involves attachment to a sand grain, shell fragment, or foraminiferan test between 0.5 and 4 mm in diameter (Cook & Chimonides 1983). Settlement on larger substrata seems to result in distorted growth or failure to develop.

There is some evidence that sexual reproduction is delayed or absent in populations undergoing asexual reproduction, just as it is in sessile species undergoing unrestrained growth (§ 5.6). Germ cells were absent in Brazilian populations of *D. u. depressa* undergoing zoarial budding (Marcus & Marcus 1962). Similarly, in populations of this species off Florida undergoing extensive fragmentation, recently settled colonies are much less common than those of *C. doma* which fragments little (Winston 1986). The relative frequency of these different processes of asexual and sexual reproduction is readily distinguished in skeletons of Recent and fossil colonies (Cook 1965a). Origin from fragments is evident from redirection of zooidal growth (Fig. 9.13), just as is regeneration of encrusting colonies (Fig. 5.16, Jackson 1983). Zoarial budding can be identified by

(a) lack of a central substratum (sand grain),
(b) reversed orientation of the regenerated zooecium that develops from the broken pseudoancestrula, and
(c) early fan-shaped budding.

In contrast, colonies derived from larvae commonly contain a central primary substratum, and bud symmetrically from an ancestrula or ancestrular region.

By virtue of its colony-wide coordination of mandibles and lophophores, mobility, ability to turn over, movement towards light, determinate growth, extreme sexual polymorphism, and zonation of feeding and sexual reproductive functions, *Selenaria maculata* is perhaps the most highly integrated bryozoan known. Furthermore, morphological evidence suggests that many other Lunulitidae are similarly highly integrated. It is of considerable interest, therefore, that no cupuladriid is likely to match this degree of integration, particularly with regard to colony agility and sexual polymorphism, despite the fact that cupuladriids have replaced (but apparently did not displace) lunulitids over much of the globe in the last 15 million years (Cook & Chimonides 1983).

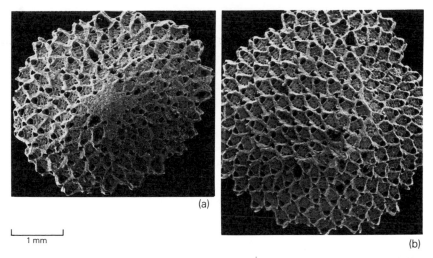

(a)

(b)

1 mm

Figure 9.13 Morphological evidence for mode of colony origin in *Discoporella umbellata*, Waccamaw Formation (Pleistocene), North Carolina: (a) regular budding pattern of colony developed from ancestrula, (b) irregular and asymmetrical pattern of colony developed from a fragment.

9.2 Rooted bryozoans

Rooted bryozoans grow in many shapes (Fig. 9.14), but the majority are erect. Some species grow 10–20 cm high, but most reach only 2–3 cm, and many are minute. The majority are cheilostomes, but there are also many ctenostomes and cyclostomes. Rooted bryozoans are the predominant form of erect bryozoans on sediments below about 1,000 m (inverse of the curve for cemented species in Fig. 4.12). They comprise an extremely heterogeneous assemblage, the great abundance and diversity of which has been realized only recently. Many groups are barely described, and our discussion is necessarily limited to a few of the more thoroughly studied forms.

The great majority of large rooted bryozoans are arborescent anascans or ctenostomes which are flexible, either because they are lightly calcified or uncalcified, or because their rigid branches are jointed at uncalcified nodes (Fig. 9.14). In addition, some rooted ascophorans form rigid branched or blade-like erect colonies 10 cm or more high. All are anchored to the bottom by kenozooidal rootlets (rhizoids) which commonly attach to a variety of objects including stones, shells, and other organisms such as hydroids and other bryozoans, as well as sediment grains. Many of these large rooted bryozoans occur in a wide range of environments and depths and are not particularly specialized for life on sediments. Many others, however, are known only from fine sediments, particularly on continental slopes or the deep sea (d'Hondt 1975, 1981, Hayward & Cook 1979). In such cases, rhizoids may be finely branched or show tubular swellings and other structure to anchor better the colony in the sediment.

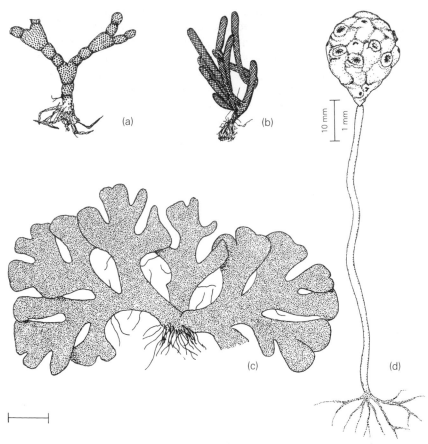

Figure 9.14 Diverse rooted gymnolaemate bryozoans: (a) *Cellarinella njegovanae* (drawn from Winston 1983, courtesy of the University of Miami Rosenstiel School of Marine Science); (b) *Cellaria* sp.; (c) *Carbasea ovoidea* (drawn after Busk 1884); (d) *Pseudalcyonidium bobinae* (drawn from d'Hondt 1975, courtesy of Université Claud Bernard, Lyon).

In contrast to the above groups are a host of bizarre cheilostomes whose adaptations to life on sediments are reflected in their entire morphology, rather than simply the presence of rootlets (Harmer 1957, Cook & Lagaaij 1976, Cook 1979, 1981, Cook & Chimonides 1981b). Most colonies are minute, ranging from 0.5 to 5 mm diameter, but some reach 3–4 cm. They are often extremely abundant, exceeding more than 20,000 colonies m^{-2}, particularly between depths of about 100–500 m (Cook 1981). Colonies consist of globular, conical, stellate, lanceolate, discoidal, or lobate masses of zooids that are supported above the sea floor or within the sediments by kenozooidal or extrazooidal rootlets, which may emanate separately from the base of the colony or from a single basal stalk. One or more rootlets arise from the ancestrula and new ones develop from younger zooids; all may lengthen and widen as the colony grows (Fig. 3.16). Rootlets are adhesive and attach to clusters of foraminiferans, small shell fragments, sand grains, worm tubes, and other minute hydrozoans and bryozoans (Fig. 9.15).

The principal anascan representatives of the specialized rooted fauna are the fan-like *Stichopora* and *Parastichopora* (Cook & Chimonides 1981b). The former is abundant in European Maastrichtian, and the latter in Recent Australian sediments. One fossil species reached 9 mm diameter, but the great majority are minute. Most other specialized rooted groups are ascophorans. Some unilaminate, discoidal species resemble lunulitids except for their development of basal rootlets on the concave surface and absence of setiform avicularia above (ibid., Cook 1979). Most other species are multilaminate, and commonly grow by the unusual "frontal astogeny" described previously in section 3.4 (Cook & Lagaaij 1976). Many common Recent genera such as *Batopora* and *Conescharellina* date from the Eocene.

Population studies of specialized rooted species are lacking except for some size-frequency data for a Cretaceous *Stichopora* (Håkansson 1975). Nevertheless, a variety of circumstantial evidence suggests a very different pattern from that of most free-living species. Truncated size-frequency distributions imply approximately determinate growth. Sexual reproduction

1 cm

Figure 9.15 Diagrammatic sketch of assemblage of rooted bryozoans (shown in silhouette), attached to worm tubes and hydroids (shown stippled). (From Cook 1979, courtesy of the British Museum (Natural History) and Academic Press.)

appears to begin very early in astogeny; in many species brood chambers occur in colonies smaller than 1 mm diameter (12 zooids) (Cook 1981). There is also no evidence for asexual reproduction, either by fragmentation or budding.

Much about development can be inferred from astogenetic series and the position of different stages in undisturbed sediments (Fig. 9.16, Cook & Chimonides 1981c). In *"Sphaeropora fossa"*, ancestrulae possess large, complex rootlets. These expand and lengthen greatly as the colony develops, reaching 10–30 mm length in adult colonies. Most of the rootlet is buried, however, with adult colonies projecting only about 5 mm above the sediment surface, as marked by a ring of secondary rootlets around the primary stalk. Colonies with rootlets shorter than about 7 mm are known to live interstitially, often attached to adults of the same species.

The possession of rootlets and the high probability of determinate growth suggest considerable colonial integration. This is further supported by the suggestion that colonies may alter their positions by differential growth of rootlets.

9.3 Interstitial bryozoans

Interstitial bryozoans live within sediments, often attached only to single grains. The ctenostomes *Monobryozoon* and *Aethozoon* grow as very large, sausage-shaped, typically solitary zooids that are anchored by adhesive pseudostolons which arise from the zooidal base (Fig. 9.17). New zooids arise from distal swellings of pseudostolons that are separated from the parental zooid by a septal partition, thereby marginally preserving the uniquely clonal (but not colonial) status of the phylum. Eventually the parental zooid dies, leaving behind its solitary clonal progeny.

Bryozoans encrusting sand grains (Fig. 9.3, § 7.1) are also interstitial by virtue of the position of their substrata within sediments. They differ from forms like *Monobryozoon*, however, in their tendency to outgrow their substrata to become free-living, or to form larger, more complex colonies on stable substrata. Some species (cryptic species complexes?) of *Cleidochasma*, for example, grow in various combinations of interstitial, circumrotatory, rooted lunulitiform, encrusting, massive, or erect colonies (Harmer 1957).

9.4 Major evolutionary trends

The only sediment-dwelling bryozoans whose habitat distribution does not appear to have changed radically throughout the Phanerozoic are circumrotatory colonies. Like similar corals and crustose algae, these are common wherever coarse bottom sediments are frequently moved about by water movements, or browsing organisms (Glynn 1974, Gill & Coates 1977, Rider & Enrico 1979, Balson & Taylor 1982).

The decline and eventual disappearance of large laminar bryozoans from the surfaces of unconsolidated sediments from the Middle to Late Paleozoic was part of the wholesale reduction of immobile animals from these environments, apparently due to dramatic increases in the intensity of

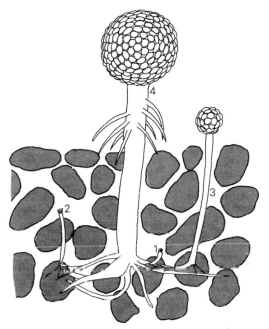

Figure 9.16 Life habit of different growth stages of rooted *"Sphaeropora fossa"*; the largest colony is nearly 15 mm tall. Successive growth stages are numbered 1–4. (Drawn after Cook & Chimonides 1981b.)

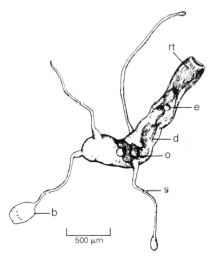

Figure 9.17 *Monobryozoon limicola*: rt, retracted tentacles; e, egg; d, digestive tract; o, ovary; s, stolon; b, bud. (From Berge *et al*. 1985, courtesy of Universitetet i Bergen.)

bioturbation (Thayer 1983). After this demise there was a long hiatus before the more-or-less simultaneous invention by many different clades of entirely new growth forms highly specialized for life on sediments. All of these specialized groups (Lunulitidae, Cupuladriidae, Setosellinidae, Mamilloporidae, Orbituliporidae, Conescharellinidae, *Stichopora*, etc.) are cheilostomes, and all appeared during the Upper Cretaceous to Eocene (Berthelsen 1962, Cheetham 1963, 1966 & 1971, Håkansson 1975, Cook & Lagaaij 1976, Cook & Chimonides 1981a, 1983). No comparable stenolaemates seem ever to have developed.

The evolution of free-living bryozoans follows closely that of comparably active free-living corals (Gill & Coates 1977). Free-living Microselenidae and Caryophyllidae first appeared in the Middle Jurassic, and free-living Fungiidae and Flabellidae during the Cretaceous (Wells 1956). The principal difference between the corals and bryozoans is that the great majority of the corals are solitary polyps whereas all the bryozoans are colonial, although their growth is quasi-determinate like most solitary corals. Moreover, the two groups are remarkably convergent in shape, size, life history, and behavior. Individuals in both groups are predominantly discoidal or domal and are small, although the corals are generally larger. For example, the median diameter of 28 Recent Mediterranean free-living corals is 14 mm (Zibrowius 1980) versus 7 mm for 27 species of Recent Lunulitidae and Cupuladriidae (Cook 1965a, b, Cadee 1975, Cook & Chimonides 1984a, b, 1985a, b, c). Thirteen of the 28 Mediterranean corals measured are clonal, and as with many populations of *Discoporella umbellata*, several species are known only from specimens regenerated from fragments (Zibrowius 1980). Lastly, all the free-living corals observed so far can exhume themselves, turn over, and move about (Goreau & Yonge 1968, Hubbard 1972, Gill & Coates 1977).

There is also striking interphyletic convergence between several minute rooted bryozoans specialized for life on sediments and benthic Foraminifera where both groups co-occur on continental slopes and in the deep sea (Cook 1981).

There is no evidence for dramatic changes in physical conditions on the sea floor at the times of these convergent radiations. It is tempting, therefore, to attribute these often contemporary and convergent radiations to increased predation pressure or bioturbation, especially for the free-living forms that are commonly buried shallowly beneath the sediment (Winston & Håkansson 1986). The deeper occurrence of rooted species compared with free-living forms is also compatible with this hypothesis, because densities of predators and bioturbators fall off considerably with increasing depth. Alternatively, they may represent the coincidental evolution among both scleractinians and cheilostomes of the morphological plasticity necessary for these bizarre patterns of growth.

0 Trends in bryozoan evolution

This chapter brings together patterns and trends in bryozoan growth forms presented in previous chapters. Our purpose is to emphasize the abundance of such patterns, their extremely long duration, and apparent adaptive significance (see also Jackson & McKinney in press).

There have also been major changes in zooidal functional morphology, but these are generally less well understood. Some zooidal patterns are related to apparent phylogenetic contraints, such as the limitation of stenolaemates to drawing water into zooids only through the orifice, and other related aspects of their hydrostatic systems, as compared with the greater diversity of hydrostatic mechanisms of gymnolaemates. Other patterns, such as changes in frequencies and types of cheilostome heteromorphic zooids, or the different modes of extending and retracting cheilostome polypides, almost certainly have adaptive significance, although this has not been demonstrated. Still other features such as zooidal size, which is apparently severely constrained by the basic bryozoan body plan, do not change. This is in striking contrast with the enormous variation that occurs in colony size and form.

10.1 Trends in the design, distribution, and relative frequency of growth forms

We have identified eight patterns, or sets of patterns, which we believe imply progressive adaptive evolution (Table 10.1). All are macroevolutionary patterns, i.e., statistical differences or trends among groups of species within many bryozoan clades. Moreover, scatter in the evolving morphologies at any one time always has been great. None of these patterns has been documented within a single clade (e.g., *Metrarabdotos*), nor is there reason to believe that they should always be apparent at that level.

The first two trends describe changes in proportions and design of erect species that can be well understood in terms of known biomechanical constraints on feeding (Fig. 6.16) and resistance of colonies to breakage (Fig. 8.11). Sampling was extensive, unbiased taxonomically, and rigorous; and the patterns clearly significant. The same is true of the third trend, in proportions of budding modes among encrusting species (Fig. 7.10), which can be clearly related to patterns of overgrowth and injury known to be important causes of mortality in these animals.

Patterns 4 and 5 involve zooidal morphology, arrangement, and integration in relation to colony form (Fig. 1.9, Table 4.1). The patterns are only suggestive, however, either because sampling was arbitrary (Cheetham, personal communication) or included too few times to detect a trend. If they

Table 10.1   Temporal trends in bryozoan evolution described in this book.

Trend	Time	Interpretation
1 increased diversity and abundance of branched unilaminate erect species relative to other forms (Ch. 6, McKinney 1986a)	Ordovician–Permian; Jurassic–Recent	lower energetic cost of feeding due to decreased resistance to outflow of filtered water; no loss of feeding surface for maculae
2 increases in (a) branch thickening rates and (b) bifurcation angles among rigidly erect adeoniform colonies (Ch. 8, Cheetham 1986a)	Cretaceous–Recent	(a) decreased chances of colony failure due to fluid or other forces, and (b) minimized crowding of branches
3 increased proportions of cheilostome species with (a) zooidal and (b) frontal budding (Ch. 7, Lidgard 1986)	Cretaceous–Recent	increased capacity for overgrowth and regeneration of injuries; greater resistance to predators
4 increased integration of zooids in cheilostome colonies (Chs. 1 & 4, Boardman & Cheetham 1973)	Cretaceous–Recent	improved design of multiserial species as predicted by growth form model
5 improved fit of zooidal elongation, spacing, and polymorphism to predictions of growth form model (Ch. 4, Coates & Jackson 1985)	Paleogene v. Recent	enhanced distinctiveness of growth forms to 'optimize' ecological attributes predicted by growth form model
6 shift of rigidly erect species into deeper water (Ch. 4)	Ordovician–Recent	response to increased predation pressure in shallow water
7 (a) decreased proportion of erect species (b) increased proportion of erect species (Ch. 4, Fig. 4.18)	Jurassic–Recent Ordovician–Permian	(a) increased predation pressure in shallow water, and (b) none
8 increased calcification of zooids associated with the (a) rise of fenestrates, and (b) sequential change in relative abundance of anascans, cribrimorphs, and ascophorans	Ordovician–Permian; Jurassic–Recent	increased predation pressure

hold up, these patterns are consistent with the adaptational predictions of the growth form model developed in Chapter 4. They are not biased taxonomically.

The sixth pattern is an apparently progressive environmental shift in the depth distribution of all taxa of rigidly erect bryozoan species throughout the

Phanerozoic (§ 4.5.2, Fig. 4.22). This has not been quantified for statistical analysis, due to problems of interpretation of paleodepths for most fossil assemblages. However, the apparently universal disappearance of rigidly erect bryozoans from back-barrier, shoal, and seagrass environments where they were formerly abundant is sufficient to substantiate the trend.

The ratio of encrusting to erect species increases markedly after the Paleozoic from about 0.1–1 (Fig. 4.18). During the Paleozoic, however, there was a smaller but significant downward trend. We believe the overall post-Paleozoic decrease in relative diversity of erect species is due to increased predation pressure, for which there is strong corroborative evidence in pattern 6. The Paleozoic increase in erect species is not understood; it may signify superiority of erect forms in the absence of severe biological disturbance (Jackson 1979a), or perhaps simply reflect the success of the fenestrates.

The final pattern of increased calcification of zooids in the Paleozoic and in the post-Paleozoic is well established but confounded, indeed defined, taxonomically (Fig. 1.14). We are confident, however, that it will hold up to quantitative analysis based only on actual measurements of skeletal material per zooid for entire faunas throughout the Phanerozoic. The probable basis, especially for cheilostomes, is increased predation.

10.2 Interpretation of trends

The patterns summarized in Table 10.1 suggest sustained, progressive evolution of numerous traits related to bryozoan growth form over hundreds of millions of years. More specifically, the proportions of species or higher bryozoan taxa displaying particular traits change over time in an apparently directed fashion that can be explained as adaptations to well-studied ecological processes. Moreover, because the trends are mostly polyphyletic and often arise by different mechanisms (e.g., the many methods of thickening of zooids or branches), they do not appear to result from taxonomic or developmental constraints. Some of the patterns were first predicted on theoretical grounds and subsequently tested comparatively; others are entirely *post hoc*. All claim generality at one level or another, and are therefore refutable by more detailed or different types of observations. All should be tested further. For example, explanation of trends 6 through 8 as due to increased predation pressure could be tested more rigorously by quantifying the frequency and extent of regenerated injuries recorded in bryozoan skeletons that are due to predators (Vermeij *et al.* 1981, Jackson 1983). Collectively, however, the data at hand already provide compelling evidence for sustained adaptive trends above the species level. These patterns are of comparable magnitude and importance among bryozoans as the increased proportion of mantle fusion and siphons in post-Paleozoic infaunal bivalves, which enables them to bury more deeply into sediments (Stanley 1968, 1975a).

Assuming these progressive trends will be upheld, there are two points of general interest that derive from Table 10.1. The first is that the trends apparently develop due to some form of species selection (McKinney 1986a) as defined by Stanley (1975b, 1979). They are therefore phenomena of Gould's (1985) second hierarchical tier of evolution but, contrary to his predictions, they are strongly directed and progressive, in some cases even across periods of mass extinction. For example, in relation to pattern 3, encrusting species with intrazooidal budding are less able to defend against overgrowth or regenerate injuries than species that do not bud intrazooidally. This suggests that intrazooidally budded species should be (other things being equal) more likely to become extinct when chances of injury or overgrowth are high. Since other things are never really equal, however, the pattern is likely to be very noisy, with some highly successful intrazooidally budding species present at all times, which is the case. Similarly, high variation characterizes the other well-quantified trends 1 and 2.

The second related point is that the existence of progressive trends strongly implies displacement of more poorly adapted species by better adapted ones (Dawkins & Krebs 1979), even though there is no direct evidence for any form of diffuse competition between them. This rather unsatisfactory assertion is justified merely on the grounds that the trends exist, for the following argument. Assume that all replacements of a species of some form A by species of some other forms $B-Z$ always occurred after and independently of the demise or decrease in A. Then we would expect a random distribution of new forms afterwards, i.e., no trends. What we observe in bryozoans, however, is that form A is replaced by form B (rather than $C-Z$) more often than expected by chance. These sustained trends suggest the action of some negative interaction between species to drive the patterns.

In summary, focus upon bryozoan growth forms rather than taxa has revealed many major features of bryozoan evolution, including several aspects not obviously correlated with growth form, such as modes of budding and patterns of life history. The growth form model developed in Chapter 4 provides a partial framework for understanding these patterns. The relatively high success of its predictions, corroborated by similar results for other phyla (Coates & Jackson 1985), strengthens our hope that our conclusions have some merit. Our approach also helps to highlight areas of ignorance, such as the functional basis, if any, of the widely different zooidal architecture of bryozoans upon which almost all bryozoan taxonomy is based. This is one of the most exciting areas for future investigation.

References

Abbott, I. A. & G. J. Hollenberg 1976. *Marine algae of California*. Stanford: Stanford University Press.

Adey, W. H. 1978. Coral reef morphogenesis: a multidimensional model. *Science* **202**, 831–7.

Ager, D. V. 1961. The epifauna of a Devonian spiriferid. *Quarterly Journal of the Geological Society, London* **117**, 1–10.

Anstey, R. L. & D. A. Delment 1972. Genetic meaning of zooecial chamber shapes in fossil bryozoans: Fourier analysis. *Science* **177**, 1000–2.

Anstey, R. L., J. F. Pachut & D. R. Prezbindowski 1976. Morphogenetic gradients in Paleozoic bryozoan colonies. *Paleobiology* **2**, 131–46.

Astrova, G. G. 1965. Morphologiya, istoriya razvitiya i sistema ordovikskikh i siluriyskikh mshanok. *Trudy Paleontologischeskogo Instituta* **106**, 1–432.

Atkins, D. 1932. The ciliary feeding mechanism of the entoproct Polyzoa, and a comparison with that of the ectoproct Polyzoa. *Quarterly Journal of Microscopical Science* **75**, 393–423.

Ayling, A. L. 1983. Growth and regeneration rates in thinly encrusting Demospongiae from temperate waters. *Biological Bulletin* **165**, 343–52.

Ayling, A. M. 1981. The role of biological disturbance in temperate subtidal encrusting communities. *Ecology* **62**, 830–47.

Balson, P. S. 1981. Facies-related distribution of bryozoans of the Coralline Crag (Pliocene) of eastern England. In *Recent and Fossil Bryozoa*, G. P. Larwood and C. Nielsen (eds.), 1–6. Fredensborg, Denmark: Olsen & Olsen.

Balson, P. S. & P. D. Taylor 1982. Palaeobiology and systematics of large cyclostome bryozoans from the Pliocene Coralline Crag of Suffolk. *Palaeontology* **25**, 529–54.

Bancroft, A. J. 1986. Secondary nanozooecia in some Upper Palaeozoic fenestrate Bryozoa. *Palaeontology* **29**, 207–12.

Banta, W. C. 1969. The body wall of cheilostome Bryozoa. II. Interzoidal communication organs. *Journal of Morphology* **129**, 149–70.

Banta, W. C. 1971. The body wall of cheilostome Bryozoa. IV. The frontal wall of *Schizoporella unicornis* (Johnston). *Journal of Morphology* **135**, 165–84.

Banta, W. C. 1972. The body wall of cheilostome Bryozoa. V. Frontal budding in *Schizoporella unicornis floridana*. *Marine Biology* **14**, 63–71.

Banta, W. C. 1975. Origin and early evolution of cheilostome Bryozoa. *Documents des Laboratoires de Géologie de la Faculté des Sciences de Lyon, Hors Série* **3**, 565–82.

Banta, W. C. & R. E. Wass 1979. Catenicellid cheilostome Bryozoa. I. Frontal walls. *Australian Journal of Zoology Supplementary Series* **68**, 1–70.

Banta, W. C., F. K. McKinney & R. L. Zimmer 1974. Bryozoan monticules: excurrent water outlets? *Science* **185**, 783–4.

Bassler, R. S. 1906. The bryozoan fauna of the Rochester Shale. *U.S. Geological Survey Bulletin* **292**, 1–137.

Bassler, R. S. 1911. *Corynotrypa*, a new genus of tubuliporoid Bryozoa. *Proceedings of the U.S. National Museum* **39**, 497–527.

Bassler, R. S. 1953. Bryozoa. In *Treatise on invertebrate paleontology, Part G*, R. C. Moore (ed.). Lawrence: University of Kansas Press.

Berge, J. A., H. P. Leinaas & K. Sandøy 1985. The solitary bryozoan, *Monobryozoan limicola*, Franzén (Ctenostomata), a comparison of mesocosm and field samples from Oslofjorden, Norway. *Sarsia* **70**, 91–4.

Bernstein, B. B. & N. Jung 1979. Selective pressures and coevolution in a kelp canopy community in southern California. *Ecology* **49**, 335–55.

Berthelsen, O. 1962. Cheilostome Bryozoa in the Danian deposits of east Denmark. *Danmarks Geologiske Undersøgelse* **83**, 1–290.

Best, B. A. & J. E. Winston 1984. Skeletal strength of encrusting cheilostome bryozoans. *Biological Bulletin* **167**, 390–409.

Best, M. A. & J. P. Thorpe 1983. Effects of particle concentration on clearance rate and feeding current velocity in the marine bryozoan *Flustrellidra hispida*. *Marine Biology* **77**, 85–92.

Best, M. A. & J. P. Thorpe 1985. Autoradiographic study of feeding and colonial transport of metabolites in the marine bryozoan *Membranipora membranacea*. *Marine Biology* **84**, 295–300.

Best, M. A. & J. P. Thorpe 1986a. Effects of food particle concentration on feeding current velocity in six species of marine Bryozoa. *Marine Biology* **93**, 255–62.

Best, M. A. & J. P. Thorpe 1986b. Feeding-current interactions and competition for food among the bryozoan epiphytes of *Fucus serratus*. *Marine Biology* **93**, 371–5.

Bigey, F. 1981. Overgrowths in Palaeozoic Bryozoa: examples from Devonian forms. In *Recent and fossil Bryozoa*, G. P. Larwood & C. Nielsen (eds.), 7–17. Fredensborg, Denmark: Olsen & Olsen.

Birkeland, C. 1977. The importance of rate of biomass accumulation in early successional stages of benthic communities to the survival of coral recruits. *Proceedings of the 3rd International Coral Reef Symposium* **1**, 15–21.

Blake, D. B. 1976. Functional morphology and taxonomy of branch dimorphism in the Paleozoic bryozoan genus *Rhabdomeson*. *Lethaia* **9**, 169–78.

Boaden, P. J. S., R. J. O'Connor & R. Seed 1975. The composition and zonation of a *Fucus serratus* community in Strangford Lough, Co. Down. *Journal of Experimental Marine Biology and Ecology* **17**, 111–36.

Boardman, R. S. 1954. Morphologic variation and mode of growth of Devonian trepostomatous Bryozoa. *Science* **120**, 322–3.

Boardman, R. S. 1960. Trepostomatous Bryozoa of the Hamilton Group of New York State. *U.S. Geological Survey Professional Paper* **340**, 1–87.

Boardman, R. S. 1968. Colony development and convergent evolution of budding pattern in "rhombotrypid" Bryozoa. *Atti della Società Italiana di Scienze Naturali e del Museo Civico di Storia Naturale de Milano* **108**, 179–84.

Boardman, R. S. 1971. Mode of growth and functional morphology of autozooids in some Recent and Paleozoic tubular Bryozoa. *Smithsonian Contributions to Paleobiology* **8**, 1–51.

Boardman, R. S. 1981. Coloniality and the origin of post-Triassic tubular bryozoans. *University of Tennessee Department of Geological Sciences Studies in Geology* **5**, 70–89.

Boardman, R. S. 1983. General features of the class Stenolaemata. In *Treatise on invertebrate paleontology, Part G* (revised), R. A. Robison (ed.), 49–137. Boulder: Geological Society of America.

Boardman, R. S. 1984. Origin of the post-Triassic Stenolaemata (Bryozoa): a taxonomic oversight. *Journal of Paleontology* **58**, 19–39.

Boardman, R. S. & A. H. Cheetham 1969. Skeletal growth, intracolony variation, and evolution in Bryozoa: a review. *Journal of Paleontology* **43**, 205–33.

Boardman, R. S. & A. H. Cheetham 1973. Degrees of colony dominance in stenolaemate and gymnolaemate Bryozoa. In *Animal colonies*, R. S. Boardman, A. H. Cheetham & W. A. Oliver Jr. (eds.), 121–220. Stroudsburg: Dowden, Hutchinson & Ross.

Boardman, R. S. & A. H. Cheetham 1987. Phylum Bryozoa. In *Fossil invertebrates*, R. S. Boardman, A. H. Cheetham & A. J. Rowell (eds.), 497–549. Oxford: Blackwell Scientific Publications.

Boardman, R. S. & F. K. McKinney 1976. Skeletal architecture and preserved organs of four-sided zooids in convergent genera of Paleozoic Trepostomata (Bryozoa). *Journal of Paleontology* **50**, 25–78.

Boardman, R. S. & F. K. McKinney 1985. Soft part characters in stenolaemate taxonomy. In *Bryozoans: Ordovician to Recent*, C. Nielsen & G. P. Larwood (eds.), 35–44. Fredensborg, Denmark: Olsen & Olsen.

Boardman, R. S. & J. Utgaard 1966. A revision of the Ordovician bryozoan genera *Monticulipora, Peronopora, Heterotrypa*, and *Dekayia*. *Journal of Paleontology* **40**, 1082–108.

Boardman, R. S., A. H. Cheetham & P. L. Cook 1983. Introduction to the Bryozoa. In *Treatise on invertebrate paleontology, Part G* (revised), R. A. Robinson (ed.), 3–48. Boulder: Geological Society of America.

Bobin, G. 1977. Interzooecial communications and the funicular system. In *Biology of bryozoans*, R. M. Woollacott & R. L. Zimmer (eds.), 307–33. New York: Academic Press.

Bonsdorff, E. & O. Vahl 1982. Food preference of the sea urchins *Echinus acutus* and *E. esculentus. Marine Behaviour Physiology* **8**, 243–8.

Borg, F. 1926. Studies on recent cyclostomatous Bryozoa. *Zoologiska Bidrag från Uppsala* **10**, 181–507.

Borg, F. 1933. A revision of the recent Heteroporidae. *Zoologiska Bidrag från Uppsala* **14**, 253–394.

Bradstock, M. & D. P. Gordon 1983. Coral-like bryozoan growths in Tasman Bay, and their protection to conserve commercial fish stocks. *New Zealand Journal of Marine and Freshwater Research* **17**, 159–63.

Brande, S. & S. S. Bretsky 1982. Avoid improper statistical analysis in bryozoans: analysis of variance is suitable for study of hierarchical variation. *Journal of Paleontology* **56**, 1207–12.

Bretsky, P. W., Jr. 1970. Upper Ordovician ecology of the central Appalachians. *Yale Peabody Museum of Natural History Bulletin* **34**, 1–150.

Brood, K. 1972. Cyclostomatous Bryozoa from the Upper Cretaceous and Danian in Scandinavia. *Stockholm Contributions in Geology* **26**, 1–464.

Brood, K. 1976. Cyclostomatous Bryozoa from the coastal waters of East Africa. *Zoologica Scripta* **5**, 277–300.

Brown, D. A. 1952. *The Tertiary cheilostomatous Polyzoa of New Zealand*. London: British Museum (Natural History).

Busk, G. 1884. Report on the Polyzoa collected by H.M.S. *Challenger* during the years 1873–76. Part I, The Cheilostomata. *Report on the Scientific Results of the Voyage of H.M.S.* Challenger – *Zoology X*, **XXX**, 1–216.

Buss, L. W. 1979. Habitat selection, directional growth and spatial refuges: why colonial animals have more hiding places. In *Biology and systematics of colonial organisms*, G. Larwood & B. R. Rosen (eds.), 459–97. London: Academic Press.

Buss, L. W. 1980a. Competitive intransitivity and size-frequency distributions of interacting populations. *Proceedings of the National Academy of Sciences. U.S.A.* **77**, 5355–9.

Buss, L. W. 1980b. Bryozoan overgrowth interactions – the interdependence of competition for space and food. *Nature* **281**, 475–7.

Buss, L. W. 1981a. Group living, competition, and the evolution of cooperation in a sessile invertebrate. *Science* **213**, 1012–14.

Buss, L. W. 1981b. Mechanisms of competition between *Onychocella alula* (Hastings) and *Antropora tincta* (Hastings) on an eastern Pacific rocky shoreline. In *Recent and fossil Bryozoa*, G. P. Larwood & C. Nielsen (eds.), 39–49. Fredensborg: Olsen & Olsen.

Buss, L. W. 1985. The uniqueness of the individual revisited. In *Population biology and evolution of clonal organisms*, J. B. C. Jackson, L. W. Buss & R. E. Cook (eds.), 467–505. New Haven: Yale University Press.

Buss, L. W. & E. W. Iverson 1981. A new genus and species of Sphaeromatidae (Crustacea: Isopoda) with experiments and observations on its reproductive biology, interspecific interactions and color polymorphisms. *Postilla* **184**, 1–23.

Buss, L. W. & J. B. C. Jackson 1981. Planktonic food availability and suspension-feeder abundance: evidence of *in situ* depletion. *Journal of Experimental Marine Biology and Ecology* **49**, 151–61.

Cadee, G. C. 1975. Lunulitiform Bryozoa from the Guyana shelf. *Netherlands Journal of Sea Research* **9**, 320–43.

Caffey, H. M. 1985. Spatial and temporal variation in settlement and recruitment of intertidal barnacles. *Ecological Monographs* **55**, 313–32.

Canu, F. and R. S. Bassler 1928. Fossil and Recent Bryozoa of the Gulf of Mexico region. *Proceedings of the U.S. National Museum* **72**, 1–199.

Canu, F. and R. S. Bassler 1933. The bryozoan fauna of the Vincentown Limesand. *U.S. National Museum Bulletin* **165**, 1–108.

Carle, K. J. and E. E. Ruppert 1983. Comparative ultrastructure of the bryozoan funiculus: a blood vessel homologue. *Zeitschrift für Zoologische Systematik und Evolutionsforschung* **21**, 181–93.

Castric, A. 1974. *Les peuplements sessile du benthos rocheux de l'archipel de Glenan, Sud Bretagne*. Unpublished doctoral thesis, Université de Paris VI.

Chamberlain, J. A., Jr. 1978. Mechanical properties of coral skeleton: compressive strength and its adaptive significance. *Paleobiology* **4**, 419–35.

Chamberlain, J. A., Jr. & R. R. Graus 1975. Water flow and hydromechanical adaptations of branched reef corals. *Bulletin of Marine Science* **25**, 112–25.

Cheetham, A. H. 1963. Late Eocene zoogeography of the eastern Gulf Coast region. *Geological Society of America Memoir* **91**, 1–113.

Cheetham, A. H. 1966. Cheilostomatous Polyzoa from the upper Bracklesham beds (Eocene) of Sussex. *Bulletin of the British Museum (Natural History) Geology* **13**, 1–115.

Cheetham, A. H. 1968. Morphology and systematics of the bryozoan genus *Metrarabdotos*. *Smithsonian Miscellaneous Collections* **153**, 1–121.

Cheetham, A. H. 1971. Functional morphology and biofacies distribution of cheilostome Bryozoa in the Danian Stage (Palcocene) of southern Scandinavia. *Smithsonian Contributions to Paleobiology* **6**, 1–87.

Cheetham, A. H. 1975. Taxonomic significance of autozooid size and shape in some early multiserial cheilostomes from the Gulf Coast of the U.S.A. *Documents des Laboratoires de Géologie de la Faculté des Sciences de Lyon, Hors Série* **3**, 547–64.

Cheetham, A. H. 1986a. Branching, biomechanics and bryozoan evolution. *Proceedings of the Royal Society, London. Series B* **228**, 151–71.

Cheetham, A. H. 1986b. Tempo of evolution in a Neogene bryozoan: rates of morphologic change within and across species boundaries. *Paleobiology* **12**, 190–202.

Cheetham, A. H. 1987. Tempo of evolution in a Neogene bryozoan: are trends in single morphologic characters misleading? *Paleobiology* **13**, 286–296.

Cheetham, A. H. & P. L. Cook 1983. General features of the class Gymnolaemata. In *Treatise on invertebrate paleontology, Part G* (revised), R. A. Robison (ed.), 138–207. Boulder: Geological Society of America.

Cheetham, A. H. & L.-A. C. Hayek 1983. Geometric consequences of branching growth in adeoniform Bryozoa. *Paleobiology* **9**, 240–60.

Cheetham, A. H. & L.-A. C. Hayek 1988. Phylogeny reconstruction in the Neogene bryozoan *Metrarabdotos:* a paleontologic evaluation of methodology. *Historical Biology* **1**, 65–83.

Cheetham, A. H. & D. M. Lorenz 1976. A vector approach to size and shape comparisons among zooids in cheilostome bryozoans. *Smithsonian Contributions to Paleobiology* **29**, 1–55.

Cheetham, A. H. & E. Thomsen 1981. Functional morphology of arborescent animals: strength and design of cheilostome bryozoan skeletons. *Paleobiology* **7**, 355–83.

Cheetham, A. H., L.-A. C. Hayek & E. Thomsen 1980. Branching structure in arborescent animals: models of relative growth. *Journal of Theoretical Biology* **85**, 335–69.

Cheetham, A. H., L.-A. C. Hayek & E. Thomsen 1981. Growth models in fossil arborescent cheilostome bryozoans. *Paleobiology* **7**, 68–86.

Chimonides, P. J. & P. L. Cook 1981. Observations on living colonies of *Selenaria* (Bryozoa, Cheilostomata). II. *Cahiers de Biologie Marine* **22**, 207–19.

Coates, A. G. & J. B. C. Jackson 1985. Morphological themes in the evolution of clonal and aclonal marine invertebrates. In *Population biology and evolution of clonal organisms*, J. B. C. Jackson, L. W. Buss & R. E. Cook (eds.), 67–106. New Haven: Yale University Press.

Cook, P. L. 1963. Observations on live lunulitiform zoaria of Polyzoa. *Cahiers de Biologie Marine* **4**, 407–13.

Cook, P. L. 1964. Polyzoa from West Africa. Notes on the genera *Hippoporina* Neviani, *Hippoporella* Canu, *Cleidochasma* Harmer and *Hippoporidra* Canu & Bassler (Cheilostomata, Ascophora). *Bulletin of the British Museum (Natural History) Zoology* **12**, 1–35.

Cook, P. L. 1965a. Notes on the Cupuladriidae (Polyzoa, Anasca). *Bulletin of the British Museum (Natural History) Zoology* **13**, 151–87.

Cook, P. L. 1965b. Polyzoa from West Africa, the Cupuladriidae (Cheilostomata, Anasca). *Bulletin of the British Museum (Natural History) Zoology* **13**, 189–227.

Cook, P. L. 1968. Observations on living Bryozoa. *Atti della Società Italiana di Scienze Naturali e del Museo Civico di Storia Naturale di Milano* **108**, 155–60.

Cook, P. L. 1973. Settlement and early colony development in some Cheilostomata. In *Living and fossil Bryozoa*, G. P. Larwood (ed.), 65–71. London: Academic Press.

Cook, P. L. 1975. The genus *Tropidozoum* Harmer. *Documents des Laboratoires de Géologie de la Faculté des Sciences de Lyon, Hors Série* **3**, 161–8.

Cook, P. L. 1977a. Colony-wide water currents in living Bryozoa. *Cahiers de Biologie Marine* **18**, 31–47.

Cook, P. L. 1977b. The genus *Tremogasterina* Canu (Bryozoa, Cheilostomata). *Bulletin of the British Museum (Natural History) Zoology* **32**, 103–65.

Cook, P. L. 1979. Mode of life of small, rooted "sand fauna" colonies of Bryozoa. In *Advances in bryozoology*, G. P. Larwood & M. B. Abbott (eds.), 269–82. London: Academic Press.

Cook, P. L. 1981. The potential of minute bryozoan colonies in the analysis of deep sea sediments. *Cahiers de Biologie Marine* **22**, 89–106.

Cook, P. L. 1986. Bryozoa from Ghana – a preliminary survey. *Koninklijk Museum voor Midden-Afrika (Tevuren, België), Zoologische Wetenschappen – Annals* **238**, 1–315.

Cook, P. L. & P. J. Chimonides 1978. Observations on living colonies of *Selenaria* (Bryozoa, Cheilostomata). I. *Cahiers de Biologie Marine* **19**, 147–58.

Cook, P. L. & P. J. Chimonides 1980. Further observations on water current patterns in living Bryozoa. *Cahiers de Biologie Marine* **21**, 393–402.

Cook, P. L. & P. J. Chimonides 1981a. Morphology and systematics of some interior-walled cheilostome Bryozoa. *Bulletin of the British Museum (Natural History) Zoology* **41**, 53–89.

Cook, P. L. & P. J. Chimonides 1981b. Morphology and systematics of some rooted cheilostome Bryozoa. *Journal of Natural History* **15**, 97–134.

Cook, P. L. & P. J. Chimonides 1981c. Early astogeny of some rooted cheilostome Bryozoa. In *Recent and fossil Bryozoa*, G. P. Larwood & C. Nielsen (eds.), 59–64. Fredensborg, Denmark: Olsen & Olsen.

Cook, P. L. & P. J. Chimonides 1983. A short history of the lunulite Bryozoa. *Bulletin of Marine Science* **33**, 566–81.

Cook, P. L. & P. J. Chimonides 1984a. Recent and fossil Lunulitidae (Bryozoa: Cheilostomata) 1. The genus *Otionella* from New Zealand. *Journal of Natural History* **18**, 227–54.

Cook, P. L. & P. J. Chimonides 1984b. Recent and fossil Lunulitidae (Bryozoa: Cheilostomata) 2. Species of *Helixotionella* gen. nov. from Australia. *Journal of Natural History* **18**, 255–70.

Cook, P. L. & P. J. Chimonides 1985a. Recent and fossil Lunulitidae (Bryozoa: Cheilostomata) 3. "Opesiulate" and other species of *Selenaria sensu lato*. *Journal of Natural History* **19**, 285–322.

Cook, P. L. & P. J. Chimonides 1985b. Recent and fossil Lunulitidae (Bryozoa: Cheilostomata) 4. American and Australian species of *Otionella*. *Journal of Natural History* **19**, 575–603.

Cook, P. L. & P. J. Chimonides 1985c. Recent and fossil Lunulitidae (Bryozoa: Cheilostomata) 5. *Selenaria alata* Tenison Woods, and related species. *Journal of Natural History* **19**, 337–58.

Cook, P. L. & P. J. Chimonides 1986. Recent and fossil Lunulitidae (Bryozoa: Cheilostomata) 6. *Lunulites sensu lato* and the genus *Lunularia* from Australasia. *Journal of Natural History* **20**, 681–705.

Cook, P. L. & R. Lagaaij 1976. Some Tertiary and Recent conescharelliniform Bryozoa. *Bulletin of the British Museum (Natural History) Zoology* **29**, 319–76.

Corrêa, D. D. 1948. A embryologia de *Bugula flabellata* (J. V. Thompson) (Bryozoa Ectoprocta). *Universidade de São Paulo Boletins da Faculdade de Filosofia, Sciências e Letras da Universidade, Zoologia* **13**, 7–71.

Cowen, R. & J. Rider 1972. Functional analysis of fenestellid bryozoan colonies. *Lethaia* **5**, 145–64.

Cuffey, R. J. 1967. Bryozoan *Tabulipora carbonaria* in Wreford megacyclothem (Lower Permian) of Kansas. *University of Kansas Paleontological Contributions, Bryozoa, Article* **1**, 1–96.

Cuffey, R. J. 1973. An improved classification, based upon numerical–taxonomic analyses, for the higher taxa of entoproct and ectoproct bryozoans. In *Living and fossil bryozoa*, G. P. Larwood (ed.), 549–64. London: Academic Press.

Cuffey, R. J. 1977. Bryozoan contributions to reefs and bioherms through time. *Studies in Geology* **4**, 181–94.

Cuffey, R. J., C. J.·Stadum & J. D. Cooper 1981. Mid-Miocene bryozoan coquinas on the Aliso Viejo Ranch, Orange County, Southern California. In *Recent and fossil Bryozoa*, G. P. Larwood & C. Nielsen (eds.), 65–72. Fredensborg, Denmark: Olsen & Olsen.

Darwin, C. 1872. *On the origin of species by means of natural selection, or the preservation of the fdvoured races in the struggle for life,* 6th edn. (with additions and corrections). London: John Murray.

Davis, G. M. 1979. The origin and evolution of the gastropod family Pomatiopsidae, with emphasis on the Mekong River Triculinae. *Academy of Natural Science Philadelphia Monograph* **20**, 1–120.

Davis, G. M. 1981. Different modes of evolution and adaptive radiation in the Pomatiopsidae (Prosobranchia: Mesogastropoda). *Malacologia* **21**, 209–62.

Dawkins, R. & J. R. Krebs 1979. Arms races between and within species. *Proceedings of the Royal Society, London. Series B* **205**, 489–511.

Day, R. W. 1977. *The ecology of settling organisms on the coral reef at Heron Island, Queensland.* Unpublished doctoral thesis, University of Sydney.

Dayton, P. K., V. Currie, T. Gerrodette, B. D. Keller, R. Rosenthal & D. V. Tresca 1984. Patch dynamics and stability of some California kelp communities. *Ecological Monographs* **54**, 253–89.

DeBurgh, M. E. & P. V. Fankboner 1979. A nutritional association between the bull kelp *Nereocystis lutkeana* and its epizoic bryozoan *Membranipora membranacea*. *Oikos* **31**, 69–72.

Dexter, R. W. 1955. Fouling organisms attached to the American lobster in Connecticut waters. *Ecology* **36**, 159–60.

Dudley, J. E. 1973. Observations on the reproduction, early larval development, and colony astogeny of *Conopeum tenuissimum* (Canu). *Chesapeake Science* **14**, 270–8.

Duncan, H. M. 1969. Bryozoans. *Geological Society of America Memoir* **114**, 345–433.

Dyer, M. F., W. G. Fry, P. D. Fry & G. J. Cranmer 1983. Benthic regions within the North Sea. *Journal of the Marine Biological Association U.K.* **63**, 683–93.

Dyrynda, P. E. J. 1981. A preliminary study of patterns of polypide generation–degeneration in marine cheilostome Bryozoa. In *Recent and fossil Bryozoa*, G. P. Larwood & C. Nielsen (eds.), 73–81. Fredensborg, Denmark: Olsen & Olsen.

Dyrynda, P. E. J. & P. E. King 1982. Sexual reproduction in *Epistomaria bursaria* (Bryozoa:

Cheilostomata), an endozooidal brooder without polypide recycling. *Journal of Zoology, London* **198**, 337–52.

Dyrynda, P. E. J. & J. S. Ryland 1982. Reproductive strategies and life histories in the cheilostome marine bryozoans *Chartella papyraccea* and *Bugula flabellata*. *Marine Biology* **71**, 241–56.

Eggleston, D. 1972a. Factors influencing the distribution of sub-littoral ectoprocts off the south of the Isle of Man (Irish Sea). *Journal of Natural History* **6**, 247–60.

Eggleston, D. 1972b. Patterns of reproduction in the marine Ectoprocta of the Isle of Man. *Journal of Natural History* **6**, 31–8.

Eldredge, N. & S. J. Gould 1972. Punctuated equilibria: an alternative to phyletic gradualism. In *Models in paleobiology*, T. J. M. Schopf (ed.), 82–115. San Francisco: Freeman, Cooper & Co.

Ettensohn, F. R., B. C. Amig, J. C. Pashin, S. F. Greb, M. C. Harris, J. C. Black, D. J. Cantrell, C. A. Smith, T. M. McMahan, A. G. Axon & G. J. McHargue 1986. Paleoecology and paleoenvironments of the bryozoan-rich Sulphur Well member, Lexington Limestone (Middle Ordovician), central Kentucky. *Southeastern Geology* **26**, 199–219.

Farmer, J. D. 1979. Morphology and function of zooecial spines in cyclostome Bryozoa: implications for paleobiology. In *Advances in bryozoology*, G. P. Larwood & M. B. Abbott (eds.), 219–46. London: Academic Press.

Fischer, J.-C. & E. Buge 1970. *Atractosoecia incrustans* (d'Orbigny) (Bryozoa Cyclostomata) espèce bathonienne symbiotique d'un Pagure. *Bulletin de la Societé Géologique de France* **12**, 126–33.

Foster, A. B. 1984. The species concept in fossil hermatypic corals: a statistical approach. *Palaeontographica Americana* **54**, 58–69.

Foster, A. B. 1985. Intracolony variation in a common reef coral and its importance for interpreting fossil species. *Journal of Paleontology* **59**, 1359–83.

Gardiner, A. R. & P. D. Taylor 1982. Computer modelling of branching growth in the bryozoan *Stomatopora*. *Nues Jahrbuch für Geologie und Paläontolgie Abhandlungen* **163**, 389–416.

Gautier, T. G. 1973. Growth in bryozoans of the order Fenestrata. In *Living and fossil Bryozoa*, G. P. Larwood (ed.), 271–4. London: Academic Press.

Gill, G. A. & A. G. Coates 1977. Mobility, growth patterns and substrate in some fossil and Recent corals. *Lethaia* **10**, 119–34.

Gleason, M. & J. B. C. Jackson in preparation. Ecology of cryptic coral reef communities. V. Distribution, abundance, and recruitment of encrusting organisms at Los Roques, Venezuela.

Glynn, P. W. 1974. Rolling stones among the Scleractinia: mobile coralliths in the Gulf of Panama. *Proceedings of the 2nd International Coral Reef Symposium* **2**, 183–98.

Gooch, J. L. & T. J. M. Schopf 1970. Population genetics of marine species of the phylum Ectoprocta. *Biological Bulletin* **138**, 138–56.

Gooch, J. L. & T. J. M. Schopf 1971. Genetic variation in the marine ectoproct *Schizoporella errata*. *Biological Bulletin* **141**, 235–46.

Gordon, D. P. 1968. Zooidal dimorphism in the polyzoan *Hippopodinella adpressa* (Busk). *Nature* **219**, 633–4.

Gordon, D. P. 1972. Biological relationships of an intertidal bryozoan population. *Journal of Natural History* **6**, 503–14.

Gordon, D. P. 1974. Microarchitecture and function of the lophophore of a marine bryozoan. *Marine Biology* **27**, 147–63.

Gordon, J. E. 1978. *Structures: or, why things don't fall down*. London: Penguin. (Reprinted 1981, New York: Da Capo Press.)

Goreau, T. F. & C. M. Yonge 1968. Coral community on muddy sand. *Nature* **217**, 421–3.

Goryunova, R. Y. & I. P. Morozova 1979. Pozdnepaleozoyskie mshanki Mongolii. *Sovmest-naya Sovetsko-Mongolskaya Paleontologicheskaya Ekspeditsiya Trudy* **9**, 1–139.

Gould, S. J. 1985. The paradox of the first tier: an agenda for paleobiology. *Paleobiology* **11**, 2–12.

Gould, S. J. & N. Eldredge 1977. Punctuated equilibria: the tempo and mode of evolution reconsidered. *Paleobiology* **3**, 115–51.

Greeley, R. 1967. Natural orientation of lunulitiform bryozoans. *Geological Society of America Bulletin* **78**, 1179–82.

Greene, C. H. & A. Schoener 1982. Succession on marine hard substrata: a fixed lottery. *Oecologia* **55**, 289–97.

Greene, C. H., A. Schoener & E. Corets 1983. Succession on marine hard substrata: the adaptive significance of solitary and colonial strategies in temperate fouling communities. *Marine Ecology Progress Series* **13**, 121–9.

Grosberg, R. K. 1981. Competitive ability influences habitat choice in marine invertebrates. *Nature* **290**, 700–2.

Guida, V. G. 1976. Sponge predation in the oyster reef community as demonstrated with *Cliona celata* Grant. *Journal of Experimental Marine Biology and Ecology* **25**, 109–22.

Gundrum, L. E. 1979. Demosponges as substrates: an example from the Pennsylvanian of North America. *Lethaia* **12**, 105–19.

Haderlie, E. C. 1974. Growth rate, depth preference and ecological succession of some sessile marine invertebrates in Monterey Harbor. *Veliger* **17**, 1–35.

Håkansson, E. 1973. Mode of growth of the Cupuladriidae (Bryozoa, Cheilostomata). In *Living and fossil Bryozoa*, G. P. Larwood (ed.), 287–98. London: Academic Press.

Håkansson, E. 1975. Population structure of colonial organisms. A paleoecological study of some free-living Cretaceous bryozoans. *Documents des Laboratoires de Géologie de la Faculté des Sciences de Lyon, Hors Série* **3**, 385–99.

Håkansson, E. & E. Thomsen 1985. Clonal propagation in fossil cheilostomes. In *Bryozoa: Ordovician to Recent,* C. Nielsen & G. P. Larwood (eds.), 345–6. Fredensborg, Denmark: Olsen & Olsen.

Håkansson, E. & J. E. Winston 1985. Interstitial bryozoans: unexpected life forms in a high energy environment. In *Bryozoa: Ordovician to Recent,* C. Nielsen & G. P. Larwood (eds.), 124–34. Fredensborg, Denmark: Olsen & Olsen.

Hall, J. 1852. Organic remains of the lower middle division of the New York system. *Natural History of New York, Part 6. Paleontology of New York* **1**.

Harmelin, J.-G. 1973. Morphological variations and ecology of the Recent cyclostome bryozoan *"Idmonea" atlantica* from the Mediterranean. In *Living and fossil Bryozoa*, G. P. Larwood (ed.), 95–106. London: Academic Press.

Harmelin, J.-G. 1975. Relations entre la forme zoariale et l'habitat chez les bryozoaires cyclostomes. Conséquences taxonomiques. *Documents des Laboratoires de Géologie de la Faculté des Sciences de Lyon, Hors Série* **3**, 369–84.

Harmelin, J.-G. 1976. Le sous-ordre des Tubuliporina (Bryozoaires Cyclostomes) en Méditerranée, écologie et systématique. *Mémoires de l'Institut Océanographique Monaco* **10**, 1–326.

Harmelin, J.-G. 1977. Bryozoaires de Iles d'Hyères: cryptofaune bryozoologique des valves vides de *Pinna nobilis* rencontrées dans les herbiers de posidonies. *Trav. Sci. Parc Nation Port-Cros* **3**, 143–57.

Harmelin, J.-G. 1985. Bryozoan dominated assemblages in Mediterranean cryptic environments. In *Bryozoa: Ordovician to Recent,* C. Nielsen & G. P. Larwood (eds.), 135–43. Fredensborg, Denmark: Olsen & Olsen.

Harmelin, J.-G., J. Vacelet & P. Vasseur 1985. Les grottes sous-marines obscures: un milieu extreme et un remarquable biotope refuge. *Tethys* **11**, 214–29.

Harmer, S. F. 1893. On the occurrence of embryonic fission in cyclostomamous Polyzoa. *Quarterly Journal of Microscopical Science (New Series)* **34**, 199–241.

Harmer, S. F. 1923. On cellularine and other Polyzoa. *Journal of the Linnean Society, Zoology* **35**, 293–360.

Harmer, S. F. 1926. The Polyzoa of the Siboga Expedition. Part II. Cheilostomata Anasca. *Siboga-Expeditie* **28**, 181–321.

Harmer, S. F. 1957. The Polyzoa of the Siboga Expedition, Part IV, Cheilostomata Ascophora. *Siboga-Expeditie* **28**, 641–1147.

Hartman, W. D. & T. F. Goreau 1970. Jamaican coralline sponges: their morphology, ecology and fossil relatives. *Symposia of the Zoological Society, London* **25**, 205–43.

Harvell, C. D. 1984a. Predator-induced defense in a marine bryozoan. *Science* **224**, 1357–9.

Harvell, C. D. 1984b. Why nudibranchs are partial predators: intracolonial variation in bryozoan palatability. *Ecology* **65**, 716–24.

Hass, H. 1948. Beitrag zur Kenntnis der Reteporiden. *Zoologica, Stuttgart* **101**, 1–138.

Hastings, A. B. 1930. On the association of a gymnoblastic hydroid (*Zanclea protecta* sp. n.) with various cheilostomatous Polyzoa from the tropical E. Pacific. *Annals and Magazine of Natural History* **5**, 552–60.

Hayward, P. J. 1980. Invertebrate epiphytes of coastal marine algae. In *The shore environment*, J. H. Price & W. F. Farnham (eds.), 761–87. London: Academic Press.

Hayward, P. J. 1985. *Ctenostome bryozoans*. London: E. J. Brill/Dr. W. Backhuys.

Hayward, P. J. & P. L. Cook 1979. The South African Museum's *Meiring Naude* cruises. Part 9. Bryozoa. *Annals of the South African Museum* **79**, 43–130.

Hayward, P. J. & P. L. Cook 1983. The South African Museum's *Meiring Naude* cruises. Part 13. Bryozoa II. *Annals of the South African Museum* **91**, 1–161.

Hayward, P. J. & P. H. Harvey 1974. The distribution of settled larvae of the bryozoans *Alcyonidium hirsutum* (Fleming) and *Alcyonidium polyoum* (Hassall) on *Fucus serratus* L. *Journal of the Marine Biological Association U. K.* **54**, 665–76.

Hayward, P. J. & J. S. Ryland 1975. Growth, reproduction and larval dispersal in *Alcyonidium hirsutum* (Fleming) and some other Bryozoa. *Pubblicazíoni della Stazione Zoologica di Napoli* **39** *(supplemento)*: 226–41.

Hayward, P. J. & J. S. Ryland 1979. *British ascophoran bryozoans*. London: Academic Press.

Hayward, P. J. & J. S. Ryland 1985. *Cyclostome bryozoans*. London: E. J. Brill/Dr. W. Backhuys.

Hayward, P. J. & P. D. Taylor 1984. Fossil and Recent Cheilostomata (Bryozoa) from the Ross Sea, Antarctica. *Journal of Natural History* **18**, 71–94.

Highsmith, R. C. 1982. Reproduction by fragmentation in corals. *Marine Ecology Progress Series* **7**, 207–26.

Hillmer, G. 1971. Bryozoen (Cyclostomata) aus dem Unter-Hauterive von Nordwestdeutsch-land. *Mitteilungen aus dem Geologisch-Paläontologischen Institut der Universität Hamburg* **40**, 5–106.

Hinds, R. W. 1975. Growth mode and homeomorphism in cyclostome Bryozoa. *Journal of Paleontology* **49**, 875–910.

Hoffmeister, J. E., K. W. Stockman & H. G. Multer 1967. Miami Limestone of Florida and its Recent Bahamian counterpart. *Geological Society of America Bulletin* **78**, 175–90.

Holme, N. A. & J. B. Wilson 1985. Faunas associated with longitudinal furrows and sand ribbons in a tide-swept area in the English channel. *Journal of the Marine Biological Association U.K.* **65**, 1051–72.

Hondt, J.-L.d' 1975. Bryozoaries cténostomes bathyaux et abyssaux de l'Atlantique nord. *Documents des Laboratoires de Géologie de la Faculté des Sciences de Lyon, Hors Série* **3**, 311–33.

Hondt, J.-L.d' 1981. Bryozoaires cheilostomes bathyaux et abyssaux provenant des cam-pagnes océanographiques américaines (1969–1972) de l'Atlantis, du Chain, et du Knorr (Woods Hole Oceanographic Institution). *Bulletin du Museum national d'Histoire naturelle, Paris*, 4th Ser. **3**, 5–71.

Hondt, J.-L.d' 1984. Bryozoaires épibiontes sur le brachiopode articulé *Gryphus vitreus* (Born, 1778) en mer Méditérranée occidentale (Corse). *Vie et Milieu* **34**, 27–33.

Hubbard, J. A. E. B. 1972. *Diaseris distorta,* an "acrobatic" coral. *Nature* **236**, 457–9.

Hughes, T. P. 1984. Population dynamics based on individual size rather than age. *American Naturalist* **123**, 778–95.

Hughes, T. P. & J. B. C. Jackson 1985. Population dynamics and life histories of foliaceous corals. *Ecological Monographs* **55**, 141–66.

Humphries, E. M. 1975. A new approach to resolving the question of speciation in smittinid bryozoans (Bryzoa Cheilostomata). *Documents des Laboratoires Géologie de la Faculté des Sciences de Lyon, Hors Série* **3**, 19–35.

Jablonski, D. & J. W. Valentine 1981. Onshore–offshore gradients in Recent eastern Pacific shelf faunas and their paleobiogeographic significance. In *Evolution today,* G. G. E. Scudder & J. L. Reveal (eds.), 441–53. Pittsburg: Carnegie-Mellon University Hunt Institute of Botany Documents.

Jackson, J. B. C. 1974. Biogeographic consequences of eurytopy and stenotopy among marine bivalves and their evolutionary significance. *American Naturalist* **105**, 541–60.

Jackson, J. B. C. 1977a. Competition on marine hard substrata: the adaptive significance of solitary and colonial strategies. *American Naturalist* **111**, 743–67.

Jackson, J. B. C. 1977b. Habitat area, colonization, and development of epibenthic community structure. In *Biology of benthic organisms,* B. F. Keegan, P. O. Ceidigh & P. J. S. Boaden (eds.), 349–58. Oxford: Pergamon Press.

Jackson, J. B. C. 1979a. Morphological strategies of sessile animals. In *Biology and systematics of colonial organisms,* G. Larwood & B. R. Rosen (eds.), 499–555. London: Academic Press.

Jackson, J. B. C. 1979b. Overgrowth competition between encrusting cheilostome ectoprocts in a Jamaican cryptic reef environment. *Journal of Animal Ecology* **48**, 805–23.

Jackson, J. B. C. 1983. Biological determinants of present and past sessile animal distributions. In *Biotic interactions in Recent and fossil benthic communities,* M. J. S. Tevesz & P. L. McCall (eds.), 39–120. New York: Plenum Publishing.

Jackson, J. B. C. 1984. Ecology of cryptic coral reef communities. III. Abundance and aggregation of encrusting organisms with particular reference to cheilostome Bryozoa. *Journal of Experimental Biology and Ecology* **75**, 37–57.

Jackson, J. B. C. 1985. Distribution and ecology of clonal and aclonal benthic invertebrates. In *Population biology and evolution of clonal organisms,* J. B. C. Jackson, L. W. Buss & R. E. Cook (eds.), 297–355. New Haven: Yale University Press.

Jackson, J. B. C. 1986. Modes of dispersal of clonal benthic invertebrates: consequences for species' distributions and genetic structure of local populations. *Bulletin of Marine Science* **39**, 588–606.

Jackson, J. B. C. & L. W. Buss 1975. Allelopathy and spatial competition among coral reef invertebrates. *Proceedings of the National Academy of Sciences. U.S.A.* **72**, 5160–3.

Jackson, J. B. C. & A. G. Coates 1986. Life cycles and evolution of clonal (modular) animals. *Philosophical Transactions of the Royal Society London. Series B* **313**, 7–22.

Jackson, J. B. C. & T. P. Hughes 1985. Adaptive strategies of coral-reef invertebrates. *American Scientist* **73**, 265–74.

Jackson, J. B. C. & K. W. Kaufmann 1987. *Diadema antillarum* was not a keystone predator in cryptic reef environments. *Science* **235**, 687–9.

Jackson, J. B. C. & F. K. McKinney in press. Ecological processes and progressive macroevolution of marine clonal benthos. In *Biotic and abiotic factors in evolution,* R. Ross & W. Allmon (eds.). Chicago: University of Chicago Press.

Jackson, J. B. C. & S. R. Palumbi 1979. Regeneration and partial predation in cryptic coral reef environments: preliminary experiments on sponges and ectoprocts. *Colloques Internationaux de Centre National de la Recherche Scientifique* **291**, 303–8.

Jackson, J. B. C. & S. P. Wertheimer 1985. Patterns of reproduction in five common species of Jamaican reef-associated bryozoans. In *Bryozoa: Ordovician to Recent,* C. Nielsen & G. P. Larwood (eds.), 161–8. Fredensborg, Denmark: Olsen & Olsen.

Jackson, J. B. C. & J. E. Winston 1981. Modular growth and longevity in bryozoans. In *Recent and fossil Bryozoa*, G. P. Larwood & C. Nielsen (eds.), 121–6. Fredensborg, Denmark: Olsen & Olsen.

Jackson, J. B. C. & J. E. Winston 1982. Ecology of cryptic coral reef communities. I. Distribution and abundance of major groups of encrusting organisms. *Journal of Experimental Marine Biology and Ecology* **57**, 135–47.

Jackson, J. B. C., M. Gleason & G. Bruno in preparation. Ecology of cryptic coral reef communities. VI. Predator abundance, prey preference, and regenerative abilities of encrusting species.

Jackson, J. B. C., T. F. Goreau & W. D. Hartman 1971. Recent brachiopod-coralline sponge communities and their paleoecological significance. *Science* **173**, 623–5.

Jackson, J. B. C., J. E. Winston & A. G. Coates 1985. Niche breadth, geographic range, and extinction of caribbean reef-associated cheilostome Bryozoa and Scleractinia. *Proceedings of the Fifth International Coral Reef Congress* **4**, 151–8.

Jebram, D. 1973. The importance of different growth directions in the Phylactolaemata and Gymnolaemata for reconstructing the phylogeny of the Bryozoa. In *Living and fossil Bryozoa*, G. P. Larwood (ed.), 565–76. London: Academic Press.

Jebram, D. 1975. Effects of different foods on *Conopeum seurati* (Canu) (Bryozoa Cheilostomata) and *Bowerbankia gracilis* Leidy (Bryozoa Ctenostomata). *Documents des Laboratoires de Géologie de la Faculté des Sciences de Lyon, Hors Série* **3**, 97–108.

Kapp, U. S. 1975. Paleoecology of Middle Ordovician stromatoporoid mounds in Vermont. *Lethaia* **8**, 195–206.

Karklins, O. L. 1969. The cryptostome Bryozoa from the Middle Ordovician Decorah Shale, Minnesota. *Minnesota Geological Survey Special Publication* no. **6**, 1–121.

Karklins, O. L. 1984. Trepostome and cystoporate bryozoans from the Lexington Limestone and the Clays Ferry Formation (Middle and Upper Ordovician) of Kentucky. *U.S. Geological Survey Professional Paper* 1066–I, 1–105.

Karklins, O. L. 1986. Chesterian (Late Mississippian) bryozoans from the upper Chainman Shale and the lowermost Ely Limestone of western Utah. *Paleontological Society Memoirs* **17**, 1–48.

Karlson, R. 1978. Predation and space utilization patterns in a marine epifaunal community. *Journal of Experimental Marine Biology and Ecology* **31**, 225–39.

Kaufmann, K. W. 1971. The form and functions of the avicularia of *Bugula* (Phylum Ectoprocta). *Postilla* **151**, 1–26.

Kay, A. M. & M. J. Keough 1981. Occupation of patches in the epifaunal communities on pier pilings and the bivalve *Pinna bicolor* at Edithburgh, South Australia. *Oecologia* **48**, 123–30.

Keough, M. J. 1983. Patterns of recruitment of sessile invertebrates in two subtidal habitats. *Journal of Experimental Marine Biology and Ecology* **66**, 213–45.

Keough, M. J. 1984a. Effects of patch size on the abundance of sessile marine invertebrates. *Ecology* **65**, 423–37.

Keough, M. J. 1984b. Dynamics of the epifauna of the bivalve *Pinna bicolor*: interactions among recruitment, predation, and competition. *Ecology* **65**, 677–88.

Keough, M. J. 1986. The distribution of a bryozoan on seagrass blades: settlement, growth, and mortality. *Ecology* **67**, 846–57.

Keough, M. J. & A. J. Butler 1983. Temporal changes in species number in an assemblage of sessile marine invertebrates. *Journal of Biogeography* **10**, 317–30.

Keough, M. J. & H. Chernoff 1987. Dispersal and population variation in the bryozoan *Bugula neritina*. *Ecology* **68**, 199–210.

Kitamura, H. & K. Hirayama 1984. Suitable food plankton for growth of a bryozoan *Bugula neritina* under laboratory conditions. *Bulletin of the Japanese Society of Scientific Fisheries* **50**, 973–7.

Kluge, G. A. 1962. *Mshanki severnykh morel SSSR*. Moscow: Akademiya Nauk. (English translation, 1975. New Delhi: Amerind.)

Kopaevich, G. V. 1984. Atlas Mshanok Ordovika, Silura í Devona Mongolii. *Sovmestnaya Sovetsko-Mongolskaya Paleontologischeskaya Ekspeditsiya Trudy* **22**, 1–164.

LaBarbera, M. 1985. Mechanisms of spatial competition of *Discinisca strigata* (Inarticulata: Brachiopoda) in the intertidal of Panama. *Biological Bulletin* **168**, 91–105.

Labracherie, M. 1973. Les assemblages de bryozoaires des sediments muebles du Golfe de Gasgogne dans la zone W-Gironde. Contribution à la connaissance de la distribution des formes de croissance zoariale. *Bulletin de l'Institut de Géologie du Bassin Aquitaine* **13**, 87–99.

Lagaaij, R. 1963. *Cupuladria canariensis* (Busk)-portrait of a bryozoan. *Palaeontology* **6**, 172–217.

Lagaaij, R. & Y. V. Gautier 1965. Bryozoan assemblages from marine sediments of the Rhone delta, France. *Micropaleontology* **11**, 39–58.

Lang, J. C. 1974. Biological zonation at the base of a reef. *American Scientist* **62**, 272–81.

Larwood, G. P. 1962. The morphology and systematics of some Cretaceous cribrimorph Polyzoa (Pelmatoporinae). *Bulletin of the British Museum (Natural History) Geology* **6**, 1–285.

Larwood, G. P. & P. D. Taylor 1979. Early structural and ecological diversification in the Bryozoa. In *Origin of major invertebrate groups*, M. R. House (ed.), 209–34. London: Academic Press.

Leversee, G. J. 1972. Flow and feeding in fan-shaped colonies of the gorgonian coral, *Leptogorgia*. *Biological Bulletin* **151**, 344–56.

Levinsen, G. M. R. 1907. Sur la regeneration total des bryozoaires. *Oversigt over det kgl. Danske Videnskabernes Selskabs Forhandlinger* **4**, 151–9.

Liddell, W. D. & C. E. Brett 1982. Skeletal overgrowths among epizoans from the Silurian (Wenlockian) Waldron Shale. *Paleobiology* **8**, 67–78.

Liddell, W. D., S. L. Ohlhorst & A. G. Coates 1984. Modern and ancient carbonate environments of Jamaica. *Sedimenta* **10**, 1–98.

Lidgard, S. 1981. Water flow, feeding, and colony form in an encrusting cheilostome. In *Recent and fossil Bryozoa*, G. P. Larwood & C. Nielsen (eds.), 135–42. Fredensborg, Denmark: Olsen & Olsen.

Lidgard, S. 1985a. Budding process and geometry in encrusting cheilostome bryozoans. In *Bryozoans: Ordovician to Recent*, C. Nielsen & G. P. Larwood (eds.), 175–82. Fredensborg, Denmark: Olsen & Olsen.

Lidgard, S. 1985b. Zooid and colony growth in encrusting cheilostome bryozoans. *Palaeontology* **28**, 255–91.

Lidgard, S. 1986. Ontogeny in animal colonies: a persistent trend in the bryozoan fossil record. *Science* **232**, 230–2.

Lidgard, S. & J. B. C. Jackson 1982. How to be an abundant encrusting bryozoan. *Abstracts with Programs Geological Society of America* **14**, 547.

Lutaud, G. 1955. Sur la ciliature du tentacule chez les bryozoaires chilostomes. *Archives de Zoologie expérimentale et Génerale, Notes et Revue* **92**, 13–19.

Lutaud, G. 1961. Contribution a l'étude du bourgeonnement et de la croissance des colonies chez *Membranipora membranacea* (Linne), Bryozoaire chilostome. *Annales de la Société royale Zoologique de Belgique* **91**, 157–300.

Lutaud, G. 1973. L'innervation du lophophore chez le bryozoaire chilostome *Electra pilosa* (L.). *Zeitschrift für Zellforschung und Mikroskopische Anatomie* **140**, 217–34.

Lutaud, G. 1983. Autozooid morphogenesis in anascan cheilostomates. In *Treatise on invertebrate paleontology, Part G* (revised), R. A. Robison (ed.), 208–37. Boulder: Geological Society of America.

Lutaud, G. 1985. Preliminary experiments on interzooidal metabolic transfer in anascan bryozoans. In *Bryozoans: Ordovician to Recent*, C. Nielsen & G. P. Larwood (eds.), 183–91. Fredensborg, Denmark: Olsen & Olsen.

McKinney, F. K. 1969. Organic structures in a Late Mississippian trepostomatous ectoproct (bryozoan). *Journal of Paleontology* **43**, 285–8.

McKinney, F. K. 1971. Trepostomatous Ectoprocta (Bryozoa) from the lower Chickamauga Group (Middle Ordovician), Wills Valley, Alabama. *Bulletins of American Paleontology* **60**, 195–337.

McKinney, F. K. 1972. Nonfenestrate Ectoprocta (Bryozoa) of the Bangor Limestone (Chester) of Alabama. *Geological Survey of Alabama Bulletin* **98**, 1–144.

McKinney, F. K. 1975. Autozooecial budding patterns in dendroid stenolaemate bryozoans. *Documents des Laboratoires de Géologie de la Faculté des Sciences de Lyon, Hors Série* **3**, 65–76.

McKinney, F. K. 1977a. Autozooecial budding patterns in dendroid Paleozoic bryozoans. *Journal of Paleontology* **51**, 303–29.

McKinney, F. K. 1977b. Functional interpretation of lyre-shaped Bryozoa. *Paleobiology* **3**, 90–7.

McKinney, F. K. 1979. Some paleoenvironments of the coiled fenestrate bryozoan *Archimedes*. In *Advances in bryozoology*, G. P. Larwood & M. B. Abbott (eds.), 321–36. London: Academic Press.

McKinney, F. K. 1980. Erect spiral growth in some living and fossil bryozoans. *Journal of Paleontology* **54**, 597–613.

McKinney, F. K. 1981a. Intercolony fusion suggests polyembryony in Paleozoic fenestrate bryozoans. *Paleobiology* **7**, 247–51.

McKinney, F. K. 1981b. Planar branch systems in colonial suspension feeders. *Paleobiology* **7**, 344–54.

McKinney, F. K. 1983. Asexual colony multiplication by fragmentation: an important mode of genet longevity in the Carboniferous bryozoan *Archimedes*. *Paleobiology* **9**, 35–43.

McKinney, F. K. 1984. Feeding currents of gymnolaemate bryozoans: better organization with higher colonial integration. *Bulletin of Marine Science* **34**, 315–19.

McKinney, F. K. 1986a. Evolution of erect marine bryozoan faunas: repeated success of unilaminate species. *American Naturalist* **128**, 795–809.

McKinney, F. K. 1986b. Historical record of erect bryozoan growth forms. *Proceedings of the Royal Society London. Series B* **228**, 133–48.

McKinney, F. K. & R. S. Boardman 1985. Zooidal biometry of Stenolaemata. In *Bryozoans: Ordovician to Recent*, C. Nielsen & G. P. Larwood (eds.), 193–203. Fredensborg, Denmark: Olsen & Olsen.

McKinney, F. K. & H. W. Gault 1980. Paleoenvironment of Late Mississippian fenestrate bryozoans, eastern United States. *Lethaia* **13**, 127–46.

McKinney, F. K. & D. M. Raup 1982. A turn in the right direction: simulation of erect spiral growth in the bryozoans *Archimedes* and *Bugula*. *Paleobiology* **8**, 101–12.

McKinney, F. K. & R. E. Wass 1981. The double helix form of branches and its relation to polymorph distribution in *Spiruluria florea* Busk. In *Recent and fossil Bryozoa*, G. P. Larwood & C. Nielsen (eds.), 159–67. Fredensborg, Denmark: Olsen & Olsen.

McKinney, F. K., M. R. A. Listokin & C. D. Phifer 1986a. Flow and polypide distribution in the cheilostome bryozoan *Bugula* and their inference in *Archimedes*. *Lethaia* **19**, 81–93.

McKinney, F. K., F. Webb & M. J. McKinney 1986b. *In situ* bryozoans in an intertidal-shallow subtidal sedimentary sequence (Middle Ordovician, southwestern Virginia). *Abstracts with Programs Geological Society of America* **18**, 254.

Marcus, E. 1939. Bryozoarios marinhos brasileiros III. *Universidade de São Paulo Boletins da Faculdade de Philosophia, Sciências e Letras, Zoologia* **3**, 111–353.

Marcus, E. & E. Marcus 1962. On some lunulitiform Bryozoa. *Universidade de São Paulo Boletins da Faculdade de Philosophia, Sciências e Letras, Zoologia* **24**, 281–312.

Marintsch, E. J. 1981. Taxonomic reevaluation of *Prasopora simulatrix* Ulrich (Bryozoa: Trepostomata). *Journal of Paleontology* **55**, 957–61.

Maturo, F. J. S., Jr. 1957. A study of the Bryozoa of Beaufort, North Carolina, and vicinity. *Journal of the Elisha Mitchell Scientific Society* **73**, 11–68.

Maturo, F. J. S., Jr. 1973. Offspring variation from known maternal stocks of *Parasmittina nitida* (Verrill). In *Living and fossil Bryozoa*, G. P. Larwood (ed.), 577–84. London: Academic Press.

Mawatari, S. F. 1975. The life history of *Membranipora serrilamella* Osburn (Bryozoa, Cheilostomata). *Bulletin of the Liberal Arts and Science Course, School of Medicine, Nihon University* no. **3**, 19–57.

Morozova, I. P. 1961. Devonskie mshanki Minusinskikh i Kuznetskoy Kotlovin. *Akademiya Nauk SSSR Paleontologischeskogo Instituta Trudy* **86**, 1–207.

Morozova, I. P. 1970. Mshanki pozdney Permi. *Akademiya Nauk SSSR Paleontologischeskogo Instituta Trudy* **122**, 1–347.

Moyano, H. I. 1972. Familia Flustridae: ensayo de redistribución de sus especies a nivel genérico. *Boletin de la Sociedad de Biologia de Concepción* **44**, 73–101.

Moyano, H. I. 1973. *Heteropora chilensis* n. sp. Nuevo heteroporido para el Pacifico sudoriental (Bryozoa Cyclostomata). *Cahiers de Biologie Marine* **14**, 79–87.

Mundy, S. P., P. D. Taylor & J. P. Thorpe 1981. A reinterpretation of phylactolaemate phylogeny. In *Recent and fossil Bryozoa*, G. P. Larwood & C. Nielsen (eds.), 185–90. Fredensborg, Denmark: Olsen & Olsen.

Muzik, K. & S. Wainwright 1977. Morphology and habitat of five Fijian sea fans. *Bulletin of Marine Science* **27**, 308–37.

Nekoroshev, V. P. 1961. Ordovikskie i Silurijskie mshanki Sibirskoj Platformy Otryad Cryptostomata. *Trudy Vsesoyuznyy Nauchno-Issledovatelskiy Geologicheskiy Institut (VSEGEI), New Series* **41**, 1–246.

Newton, G. B. 1971. *Rhabdomesid bryozoans of the Wreford megacyclothem (Wolfcamplan, Permian) of Nebraska, Kansas, and Oklahoma*. University of Kansas Paleontological Contributions no. **56**, 1–71.

Nielsen, C. 1970. On metamorphosis and ancestrula formation in cyclostomatous bryozoans. *Ophelia* **7**, 217–56.

Nielsen, C. 1981. On morphology and reproduction of *"Hippodiplosia" insculpta* and *Fenestrulina malussi* (Bryozoa, Cheilostomata). *Ophelia* **20**, 91–125.

Nielsen, C. & K. J. Pedersen 1979. Cystid structure and protrusion of the polypide in *Crisia* (Bryozoa, Cyclostomata). *Acta Zoologia (Stockholm)* **60**, 65–88.

North, W. J. 1961. Life-span of the fronds of the giant kelp, *Macrocystis pyrifera*. *Nature* **190**, 1214–15.

Nye, O. B., Jr. 1976. Generic revision and skeletal morphology of some cerioporid cyclostomes (Bryozoa). *Bulletins of American Paleontology* **69**, 1–222.

O'Connor, R. J., R. J. Seed & P. J. S. Boaden 1980. Resource space partitioning by the Bryozoa of a *Fucus serratus* L. community. *Journal of Experimental Marine Biology and Ecology* **45**, 117–37.

Okamura, B. 1985. The effects of ambient flow velocity, colony size, and upstream colonies on the feeding success of Bryozoa. Part I. *Bugula stolonifera* Ryland, an arborescent species. *Journal of Experimental Marine Biology and Ecology* **83**, 179–93.

Osborne, S. 1984. Bryozoan interactions: observations on stolonal outgrowths. *Australian Journal of Marine and Freshwater Research* **35**, 453–62.

Osburn, R. C. 1950. Bryozoa of the Pacific coast of America. Part 1, Cheilostomata – Anasca. *Allan Hancock Pacific Expeditions* **14**, 1–269. Los Angeles: University of Southern California Press.

Osburn, R. C. 1952. Bryozoa of the Pacific coast of America. Part 2, Cheilostomata – Ascophora. *Allan Hancock Pacific Expeditions* **14**, 279–611. Los Angeles: University of Southern California Press.

Osman, R. W. 1977. The establishment and development of a marine epifaunal community. *Ecological Monographs* **47**, 37–63.

Osman, R. W. & J. A. Haugsness 1981. Mutualism among sessile invertebrates. *Science* **211**, 846–8.

Pachut, J. F. & R. L. Anstey 1984. The relative information content of Fourier-structural, binary (presence–absence) and combined data sets: a test using the H. A. Nicholson collection of Paleozoic stenolaemate bryozoans. *Journal of Paleontology* **58**, 1296–311.

Palmer, A. R. 1981. Do carbonate skeletons limit the rate of body growth? *Nature* **292**, 150–2.

Palmer, T. J. 1982. Cambrian to Cretaceous changes in hardground communities. *Lethaia* **15**, 309–23.

Palmer, T. J. & C. D. Hancock 1973. Symbiotic relationships between ectoprocts and gastropods, and ectoprocts and hermit crabs in the French Jurassic. *Palaeontology* **16**, 563–6.

Palmer, T. J. & C. D. Palmer 1977. Faunal distribution and colonization strategy in a Middle Ordovician hardground community. *Lethaia* **10**, 179–99.

Palumbi, S. R. & J. B. C. Jackson 1982. Ecology of cryptic coral reef communities. II. Recovery from small disturbance events by encrusting Bryozoa; the influence of "host" species and lesion size. *Journal of Experimental Marine Biology and Ecology* **64**, 103–15.

Palumbi, S. R. & J. B. C. Jackson 1983. Ageing in modular organisms: ecology of zooid senescence in *Steginoporella* sp. (Bryozoa: Cheilostomata). *Biological Bulletin* **164**, 267–78.

Pohowsky, R. A. 1973. A Jurassic cheilostome from England. In *Living and fossil Bryozoa*, G. P. Larwood (ed.), 447–61. London: Academic Press.

Pohowsky, R. A. 1978. The boring ctenostomate Bryozoa: taxonomy and paleobiology based on cavities in calcareous substrata. *Bulletins of American Paleontology* **73**, 1–192.

Powell, N. A. 1968. Bryozoa (Polyzoa) of Arctic Canada. *Journal of the Fisheries Research Board of Canada* **25**, 2269–320.

Probert, P. K. & E. J. Batham 1979. Epibenthic macrofauna off southeastern New Zealand and mid-shelf bryozoan dominance. *New Zealand Journal of Marine and Freshwater Research* **13**, 379–92.

Rees, J. T. 1972. The effect of current on growth form in an octocoral. *Journal of Experimental Marine Biology and Ecology* **10**, 115–23.

Rider, J. & R. Enrico 1979. Structural and functional adaptations of mobile anascan ectoproct colonies (ectoproctaliths). In *Advances in bryozoology*, G. P. Larwood & M. B. Abbott (eds.), 297–320. London: Academic Press.

Ristedt, H. & H. Schuhmacher 1985. The bryozoan *Rhynchozoon larreyi* (Audouin; 1826) – a successful competitor in coral reef communities of the Red Sea. *Marine Ecology* **6**, 167–79.

Ross, J. P. 1967. Evolution of ectoproct genus *Prasopora* in Trentonian time (Middle Ordovician) in northern and central United States. *Journal of Paleontology* **41**, 403–16.

Ross, J. P. 1970. Distribution, paleoecology, and correlation of Champlainian Ectoprocta (Bryozoa), New York State, Part III. *Journal of Paleontology* **44**, 346–82.

Ross, J. P. 1972. Paleoecology of Middle Ordovician ectoproct assemblages. *24th International Geology Congress Section* **7**, 96–102.

Rubin, J. A. 1982. The degree of intransitivity and its measurement in an assemblage of encrusting cheilostome Bryozoa. *Journal of Experimental Marine Biology and Ecology* **60**, 119–28.

Rubin, J. A. 1985. Mortality and avoidance of competitive overgrowth in encrusting Bryozoa. *Marine Ecology Progress Series* **23**, 291–9.

Rucker, J. B. 1967. Paleoecological analysis of cheilostome Bryozoa from Venezuela-British Guiana shelf sediments. *Bulletin of Marine Science* **17**, 787–839.

Russ, G. R. 1982. Overgrowth in a marine epifaunal community: competitive hierarchies and competitive networks. *Oecologia* **53**, 12–19.

Ryland, J. S. 1962. The association between Polyzoa and algal substrata. *Journal of Animal Ecology* **31**, 331–8.

Ryland, J. S. 1963. Systematic and biological studies on Polyzoa (Bryozoa) from western Norway. *Sarsia* **14**, 1–59.

Ryland, J. S. 1965. *Catalogue of Main Marine Fouling Organisms*, Vol. 3, 1–83. Paris: Organisation for Economic Co-operation and Development.

Ryland, J. S. 1970. *Bryozoans*. London: Hutchinson University Library.

Ryland, J. S. 1974. Behaviour, settlement and metamorphosis of bryozoan larvae: a review. *Thalassia Jugoslavica* **10**, 239–62.

Ryland, J. S. 1975. Parameters of the lophophore in relation to population structure in a bryozoan community. In *Proceedings 9th European Marine Biology Symposium*, H. Barnes (ed.), 363–93. Aberdeen: Aberdeen University Press.

Ryland, J. S. 1976. Physiology and ecology of marine bryozoans. *Advances in Marine Biology* **14**, 285–443.

Ryland, J. S. 1979a. *Celleporella carolinensis* sp. nov. (Bryozoa Cheilostomata) from the Atlantic coast of America. In *Advances in bryozoology*, G. P. Larwood & M. B. Abbott (eds.), 611–20. London: Academic Press.

Ryland: J. S. 1979b. Structural and physiological aspects of coloniality in Bryozoa. In *Biology and systematics of colonial organisms*, G. Larwood & B. R. Rosen (eds.), 211–42. London: Academic Press.

Ryland, J. S. 1981. Colonies, growth and reproduction. In *Recent and fossil Bryozoa*, G. P. Larwood & C. Nielsen (eds.), 221–6. Fredensborg, Denmark: Olsen & Olsen.

Ryland: J. S. & P. J. Hayward 1977. *British anascan bryozoans*. London: Academic Press.

Ryland, J. S. & A. R. D. Stebbing 1971. Settlement and orientated growth in epiphytic and epizooic bryozoans. In *Proceedings 4th European Marine Biology Symposium*, D. J. Crisp (ed.), 105–23. London: Cambridge University Press.

Sammarco, P. W. 1980. *Diadema* and its relationship to coral spat mortality: grazing, competition, and biological disturbance. *Journal of Experimental Marine Biology and Ecology* **45**, 245–72.

Schoener, A. & T. W. Schoener 1981. The dynamics of the species–area relation in marine fouling systems. 1. Biological correlates of changes in the species–area slope. *American Naturalist* **118**, 339–60.

Schopf, T. J. M. 1969. Paleoecology of ectoprocts (bryozoans). *Journal of Paleontology* **43**, 234–44.

Schopf, T. J. M. 1974. Survey of genetic differentiation in a coastal zone invertebrate: the ectoproct *Schizoporella errata*. *Biological Bulletin* **146**, 78–87.

Schopf, T. J. M. & A. R. Dutton 1976. Parallel clines in morphologic and genetic differentiation in a coastal zone marine invertebrate: the bryozoan *Schizoporella errata*. *Paleobiology* **2**, 255–64.

Schopf, T. J. M. & J. L. Gooch 1971. Gene frequencies in a marine ectoproct: a cline in natural populations related to sea temperature. *Evolution* **25**, 286–9.

Schopf, T. J. M., K. O. Collier & B. O. Bach 1980. Relation of the morphology of stick-like bryozoans at Friday Harbor, Washington, to bottom currents, suspended matter and depth. *Paleobiology* **6**, 466–76.

Seed, R. & S. Harris 1980. The epifauna of the fronds of *Laminaria digitata* Lamour in Strangford Lough, Northern Ireland. *Proceedings of the Royal Irish Academy* **80B**, 91–106.

Seed, R. & R. J. O'Connor 1981. Community organization in marine algal epifaunas. *Annual Review of Ecology and Systematics* **12**, 49–74.

Sibley, C. G. & J. E. Ahlquist 1983. Phylogeny and classification of birds based on the data of DNA–DNA hybridization. In *Current ornithology*, Volume 1, R. F. Johnston (ed.), 249–92. New York: Plenum Press.

Silén, L. 1938. Zur Kenntnis des Polymorphismus der Bryozoen. Die Avicularien der Cheilostomata Anasca. *Zoologiska Bidrag från Uppsala* **17**, 149–366.

Silén, L. 1966. On the fertilization problem in gymnolaematous Bryozoa. *Ophelia* **3**, 113–40.

Silén, L. 1972. Fertilization in the Bryozoa. *Ophelia* **10**, 27–34.

Silén, L. 1977. Polymorphism. In *Biology of bryozoans,* R. M. Woollacott & R. L. Zimmer (eds.), 183–231. New York: Academic Press.

Silén, L. & J.-G. Harmelin 1974. Observations on living Diastoporidae (Bryozoa Cyclostomata), with special regard to polymorphism. *Acta Zoologia* **55,** 81–96.

Simonsen, A. H. & R. J. Cuffey 1980. Fenestrate, pinnate, and ctenostome bryozoans and associated barnacle borings in the Wreford megacyclothem (Lower Permian) of Kansas, Oklahoma, and Nebraska. *University of Kansas Paleontological Contributions no.* **101,** 1–38.

Smith, D. B. 1981. Bryozoan-algal patch-reefs in the Upper Permian Lower Magnesian Limestone of Yorkshire, Northeast England. *Society of Economic Paleontologists and Mineralogists Special Publication* **30,** 187–202.

Smith, L. W. 1973. Ultrastructure of the tentacles of *Flustrellidra hispida* (Fabricius). In *Living and fossil Bryozoa,* G. P. Larwood (ed.), 335–42. London: Academic Press.

Sokal, R. R. 1985. The continuing search for order. *American Naturalist* **126,** 729–49.

Soule, D. F. 1973. Morphogenesis of giant avicularia and ovicells in some Pacific Smittinidae. In *Living and fossil Bryozoa,* G. P. Larwood (ed.), 485–95. London: Academic Press.

Soule, D. F. & J. D. Soule 1973. Morphology and speciation of Hawaiian and eastern Pacific Smittinidae (Bryozoa, Ectoprocta). *Bulletin of the American Museum of Natural History* **152,** 369–440.

Sparling, D. R. 1964. *Prasopora* in a core from the Northville area, Michigan. *Journal of Paleontology* **38,** 1072–81.

Spjeldnaes, N. 1975. Silurian bryozoans which grew in the shade. *Documents des Laboratoires de Géologie de la Faculté des Sciences de Lyon, Hors Série* **3,** 415–24.

Stach, L. W. 1937. The application of Bryozoa in Cainozoic Stratigraphy. *Australia and New Zealand Association for the Advancement of Science Report of the 23rd Meeting* 80–83.

Stanley, S. M. 1968. Post-Paleozoic adaptive radiation of infaunal bivalve molluscs – a consequence of mantle fusion and siphon formation. *Journal of Paleontology* **42,** 214–29.

Stanley, S. M. 1975a. Adaptive themes in the evolution of the Bivalvia (Mollusca). *Annual Review of Earth and Planetary Science* **3,** 361–85.

Stanley, S. M. 1975b. A theory of evolution above the species level. *Proceedings of the National Academy of Sciences U.S.A.* **72,** 646–50.

Stanley, S. M. 1979. *Macroevolution: pattern and process.* San Francisco: W. H. Freeman.

Stebbing, A. R. D. 1971. Growth of *Flustra foliacea* (Bryozoa). *Marine Biology* **9,** 267–73.

Stebbing, A. R. D. 1972. Preferential settlement of a bryozoan and serpulid larvae on the younger parts of *Laminaria* fronds. *Journal of the Marine Biology Association U.K.* **52,** 765–72.

Stebbing, A. R. D. 1973a. Competition for space between the epiphytes of *Fucus serratus* L. *Journal of the Marine Biology Association U.K.* **53,** 247–61.

Stebbing, A. R. D. 1973b. Observations on colony overgrowth and spatial competition. In *Living and fossil Bryozoa,* G. P. Larwood (ed.), 173–83. London: Academic Press.

Steneck, R. S. 1983. Escalating herbivory and resulting adaptive trends in calcareous algal crusts. *Paleobiology* **9,** 44–61.

Stimson, J. 1985. The effect of shading by the table coral *Acropora hyacinthus* on understory corals. *Ecology* **66,** 40–53.

Strathmann, R. R. 1973. Function of lateral cilia in suspension feeding of lophophorates (Brachiopoda, Phoronida, Ectoprocta). *Marine Biology* **23,** 129–36.

Strathmann, R. R. 1982. Cinefilms of particle capture by an induced local change of beat of lateral cilia of a bryozoan. *Journal of Experimental Marine Biology and Ecology* **62,** 225–36.

Strom, R. 1977. Brooding patterns of bryozoans. In *Biology of bryozoans,* R. M. Woollacott & R. L. Zimmer (eds.), 23–55. New York: Academic Press.

Sutherland, J. P. 1978. Functional roles of *Schizoporella* and *Styela* in the fouling community at Beaufort, North Carolina. *Ecology* **59,** 257–64.

Sutherland, J. P. 1981. The fouling community at Beaufort, North Carolina: a study in stability. *American Naturalist* **118,** 499–519.

Sutherland, J. P. & R. H. Karlson 1977. Development and stability of the fouling community at Beaufort, North Carolina. *Ecological Monographs* **47**, 425–46.

Sverdrup, H. U., M. W. Johnson & R. H. Fleming 1942. *The Oceans.* Englewood Cliffs, N.J.: Prentice-Hall.

Tavener-Smith, R. & A. Williams 1972. The secretion and structure of the skeleton of living and fossil Bryozoa. *Philosophical Transactions of the Royal Society London. Series B* **264**, 97–159.

Taylor, P. D. 1975. Monticules in a Jurassic cyclostomatous bryozoan. *Geological Magazine* **112**, 601–6.

Taylor, P. D. 1976. Multilamellar growth in two Jurassic cyclostomatous Bryozoa. *Palaeontology* **19**, 293–306.

Taylor, P. D. 1979a. Functional significance of contrasting colony form in two Mesozoic encrusting bryozoans. *Palaegeography Palaeoclimatology Palaeoecology* **26**, 151–8.

Taylor, P. D. 1979b. Palaeoecology of the encrusting epifauna of some British Jurassic bivalves. *Palaeogeography Palaeoclimatology Palaeoecology* **28**, 241–62.

Taylor, P. D. 1979c. The inference of extrazooidal feeding currents in fossil bryozoan colonies. *Lethaia* **12**, 47–56.

Taylor, P. D. 1981. Functional morphology and evolutionary significance of differing modes of tentacle eversion in marine bryozoans. In *Recent and fossil Bryozoa,* G. P. Larwood & C. Nielsen (eds.), 235–47. Fredensborg, Denmark: Olsen & Olsen.

Taylor, P. D. 1984. Adaptations for spatial competition and utilization in Silurian encrusting bryozoans. *Special Papers in Palaeontology* **32**, 197–210.

Taylor, P. D. 1985a. Carboniferous and Permian species of the cyclostome bryozoan *Corynotrypa* Bassler, 1911 and their clonal propagation. *Bulletin of the British Museum (Natural History) Geology* **38**, 359–72.

Taylor, P. D. 1985b. Polymorphism in meliceritid cyclostomes. In *Bryozoa: Ordovician to Recent,* C. Nielsen & G. P. Larwood (eds.), 311–18. Fredensborg, Denmark: Olsen & Olsen.

Taylor, P. D., G. P. Larwood & P. S. Balson 1981. Some British field localities with fossil Bryozoa. In *Recent and fossil Bryozoa,* G. P. Larwood & C. Nielsen (eds.), 249–62. Fredensborg, Denmark: Olsen & Olsen.

Thayer, C. W. 1979. Biological bulldozers and the evolution of marine benthic communities. *Science* **203**, 458–61.

Thayer, C. W. 1983. Sediment-mediated biological disturbance and the evolution of marine benthos. In *Biotic interactions in Recent and fossil benthic communities,* M. J. S. Tevesz & P. L. McCall (eds.), 479–625. New York: Plenum Press.

Thomsen, E. 1976. Depositional environment and development of Danian bryozoan biomicrite mounds (Karlby Klint, Denmark). *Sedimentology* **23**, 485–509.

Thorpe, J. P. & J. S. Ryland 1979. Cryptic speciation detected by biochemical genetics in three ecologically important intertidal bryozoans. *Estuarine and Coastal Marine Science* **8**, 395–8.

Thorpe, J. P., J. A. Beardmore & J. S. Ryland 1978a. Genetic evidence for cryptic speciation in the marine bryozoan *Alcyonidium gelatinosum. Marine Biology* **49**, 27–32.

Thorpe, J. P., J. A. Beardmore & J. S. Ryland 1978b. Taxonomy, interspecific variation and genetic distance in the phylum Bryozoa. In *Marine organisms: genetics, ecology, and evolution,* B. Battaglia & J. A. Beardmore (eds.), 425–45. New York: Plenum.

Thorpe, J. P., J. S. Ryland & J. A. Beardmore 1978c. Genetic variation and biochemical systematics in the marine bryozoan *Alcyonidium mytili. Marine Biology* **49**, 343–50.

Thorson, G. 1950. Reproductive and larval ecology of marine bottom invertebrates. *Biological Reviews* **25**, 1–45.

Tillier, S. 1975. Recherches sur la structure et révision systématique des heteroporides (Bryozoa, Cyclostomata) des Faluns de Touraine. *Travaux du Laboratoire de Paleontologie, Université de Paris Faculté des Sciences d'Orsay.*

Ulrich, E. O. 1893. On Lower Silurian Bryozoa of Minnesota. *Minnesota Geology and Natural History Survey Final Report* no. **3**, 96–332.

Utgaard, J. 1973. Mode of colony growth, autozooids, and polymorphism in the bryozoan order Cystoporata. In *Animal colonies*, R. S. Boardman, A. H. Cheetham & W. A. Oliver, Jr. (eds.), 317–60. Stroudsburg: Dowden, Hutchinson & Ross.

Vance, R. R. 1979. Effects of grazing by the sea urchin, *Centrostephanus coronatus*, on prey community composition. *Ecology* **60**, 537–46.

Vavra, N. 1974. Cyclostome Bryozoen aus dem Badenien (Mittel miozan) von Baden bei Wien (Niederosterreich). *Neues Jahrbuch für Geologie und Paläontologie Abhandlungen* **147**, 343–75.

Vavra, N. 1983. Bryozoen aus dem Unteren Meeressand (Mitteloligozan) von Eckelsheim (Mainzer Becken, Bundesrepublik Deutschland). *Mainzer Naturwissenschaftliches Archiv* **21**, 67–123.

Velimirov, B. 1973. Orientation in the sea fan *Eunicella cavoliniee* related to water movement. *Helgoländer Wissenschaftliche Meersuntersuchungen* **24**, 163–73.

Vermeij, G. J. 1977. The Mesozoic marine revolution: evidence from snails, predators and grazers. *Paleobiology* **3**, 245–58.

Vermeij, G. J., D. E. Shindel & E. Zipser 1981. Predation through geological time: evidence from gastropod shell repair. *Science* **214**, 1024–6.

Vogel, S. 1981. *Life in moving fluids. The physical biology of flow.* Boston: Willard Grant Press.

Voigt, E. 1973. Environmental conditions of bryozoan ecology of the hardground biotope of the Maastrichtian Tuff Chalk near Maastricht (Netherlands). In *Living and fossil Bryozoa*, G. P. Larwood (ed.), 185–97. London: Academic Press.

Voigt, E. 1979. The preservation of slightly or non-calcified Bryozoa (Ctenostomata and Cheilostomata) by bioimmuration. In *Advances in bryozoology*, G. P. Larwood & M. B. Abbott (eds.), 541–64. London: Academic Press.

Voigt, E. 1981. Upper Cretaceous bryozoan–seagrass association in the Maastrichtian of the Netherlands. In *Recent and fossil Bryozoa*, G. P. Larwood & C. Nielsen (eds.), 281–98. Fredensborg, Denmark: Olsen & Olsen.

Wainwright, S. A. & J. R. Dillon 1969. On the orientation of sea fans (genus *Gorgonia*). *Biological Bulletin* **136**, 130–9.

Wainwright, S. A., W. D. Biggs, J. D. Currey & J. M. Gosline 1976. *Mechanical design in organisms*. New York: Halsted Press.

Walker, K. R. & K. F. Ferrigno 1973. Major Middle Ordovician reef tract in east Tennessee. *American Journal of Science* **273-A**, 294–325.

Walter, B. 1969. Les bryozoaires jurassiques en France. *Documents des Laboratoires de Géologie de la Faculté des Sciences de Lyon* **35**, 1–327.

Warner, D. J. & R. J. Cuffey 1973. Fistuliporacean bryozoans of the Wreford megacyclothem (Lower Permian) of Kansas. *University of Kansas Paleontological Contributions* **65**, 1–24.

Wells, H. W., M. J. Wells & I. E. Gray 1964. The calico scallop community in North Carolina. *Bulletin of Marine Science Gulf and Caribbean* **14**, 561–93.

Wells, J. W. 1956. Scleractinia. In *Treatise on Invertebrate Paleontology, Part F, Coelenterata*, R. C. Moore (ed.), 328–444. Boulder: Geological Society of America.

Wilson, J. L. 1975. *Carbonate facies in geologic history*. Berlin: Springer-Verlag.

Winston, J. E. 1976. Experimental culture of the estuarine ectoproct *Conopeum tenuissimum* from Chesapeake Bay. *Biological Bulletin* **150**, 318–35.

Winston, J. E. 1977a. Distribution and ecology of estuarine ectoprocts: a critical review. *Chesapeake Science* **18**, 34–57.

Winston, J. E. 1977b. Feeding in marine bryozoans. In *Biology of bryozoans*, R. M. Woollacott & R. L. Zimmer (eds.), 233–71. New York: Academic Press.

Winston, J. E. 1978. Polypide morphology and feeding behavior in marine ectoprocts. *Bulletin of Marine Science* **28**, 1–31.

Winston, J. E. 1979. Current-related morphology and behavior in some Pacific coast bryozoans. In *Advances in bryozoology,* G. P. Larwood & M. B. Abbott (eds.), 247–68. London: Academic Press.

Winston, J. E. 1981. Feeding behavior of modern bryozoans. *University of Tennessee Department of Geological Science. Studies in Geology no. 5,* 1–21.

Winston, J. E. 1982. Marine bryozoans (Ectoprocta) of the Indian River area (Florida). *Bulletin of the American Museum of Natural History* **173**, 99–176.

Winston, J. E. 1983. Patterns of growth, reproduction and mortality in bryozoans from the Ross Sea, Antarctica. *Bulletin of Marine Science* **33**, 688–702.

Winston, J. E. 1984a. Shallow-water bryozoans of Carrie Bow Cay, Belize. *American Museum Novitates* **2799**, 1–38.

Winston, J. E. 1984b. Why bryozoans have avicularia – a review of the evidence. *American Museum Novitates* **2789**, 1–26.

Winston, J. E. 1985. Life history studies of *Disporella* and *Drepanophora* in Jamaica. In *Bryozoa: Ordovician to Recent,* C. Nielsen & G. P. Larwood (eds.), 350. Fredensborg, Denmark: Olsen & Olsen.

Winston, J. E. 1986. *Life histories of lunulitiform bryozoans.* Final Report, National Geographic Society.

Winston, J. E. & J. B. C. Jackson 1984. Ecology of cryptic coral reef communities. IV. Community development and life histories of encrusting cheilostome Bryozoa. *Journal of Experimental Marine Biology and Ecology* **76**, 1–21.

Winston, J. E. & E. Håkansson 1986. The interstitial bryozoan fauna from Capron Shoals, Florida. *American Museum Novitates* **2865**, 1–98.

Winston, J. E. & E. Håkansson in preparation. Getting out from under: molting by *Cupuladria doma,* a free-living bryozoan.

Wood, V. & R. Seed 1980. The effects of shore level on the epifaunal communities associated with *Fucus striatus* (L) in the Menai Strait, North Wales. *Cahiers de Biologie Marine* **21**, 135–54.

Woollacott, R. M. & W. J. North 1971. Bryozoans of California and northern Mexico kelp beds. *Nova Hedwigia Beiheft* **32**, 445–79.

Woollacott, R. M. & R. L. Zimmer 1972. Origin and structure of the brood chamber in *Bugula neritina* (Bryozoa). *Marine Biology* **16**, 165–70.

Yoshioka, P. M. 1982a. Role of planktonic and benthic factors in the population dynamics of the bryozoan *Membranipora membranacea. Ecology* **63**, 457–68.

Yoshioka, P. M. 1982b. Predator-induced polymorphism in the bryozoan *Membranipora membranacea. Journal of Experimental Marine Biology and Ecology* **61**, 233–42.

Zavodnik, D. 1967. Contribution to the ecology of *Pinna nobilis* L. (Moll., Bivalvia) in the northern Adriatic. *Thalassia Jugoslavica* **3**, 93–102.

Zibrowius, H. 1980. Les scleractiniaires de la Méditerranée et de l'Atlantique nord-orientale. *Mémoirs de l'Institut Océanographique Monaco* **11**, 1–284.

Zimmer, R. L. & R. M. Woollacott 1977. Metamorphosis, ancestrulae, and coloniality in bryozoan life cycles. In *Biology of bryozoans,* R. M. Woollacott & R. L. Zimmer (eds.), 91–142. New York: Academic Press.

Index

Numbers in **bold** type refer to text sections and numbers in *italic* type refer to text figures.

Acanthotrypina 1.19
alconal animals 100, *5.5*, Table 7.1
Acropora 188
'*Actinostoma*' *6.19*
adaptive morphology 58
Adeonella 138
adeoniform colonies 33, 67, **3.2.2**, **8.3.2**, *3.4, 6.14*, Table 8.1
Aequipecten Table 7.2
Aetea 112, 131
Aethozoon 206
Aglaophenia 180
Alcyonidium 25, 38, 40, 171, *5.2*, Table 7.2
allele frequency 38, 40, *2.3, 2.4*
Amathia 131
ambient flow 84, 178–80, 186, *8.15*
Amplexopora 1.18, 1.19
Anaphragma 1.19
Anasca 30, 74, *1.1, 1.14, 1.25, 4.2, 4.3*, Table 4.2
ancestrula 17, 69, 100, 106, 177, 200, 202, 206, *5.4, 9.13*
Antropora 96, 157, *7.9*, Table 7.2
aperture 17, 33, *1.2*
Archimedes 64, 65, 94, 143, 176–7, 185, *3.12, 3.14, 8.9, 8.10*
Arthrophragma 1.20, 6.10
articulation 172
Ascophora 30, 74, *1.14, 1.25, 4.2, 4.3*
ascus 29, 51
astogeny
  conescharellinid *3.16*
  early *1.23, 5.4, 9.12*
  frontal 205, **3.4**
Atactotoechus 33
attachment and construction, erect colonies 83–4
attachment region 14–16
autozooid 9, 14, 22, **1.1.1**, *1.1, 1.2, 1.6, 1.8, 1.12*
avicularium 6–7, 9, 25, 33, 77, *2.4*
  adventitious 7, 9, 38–40, 44, *1.2, 1.3, 1.4, 1.6, 2.2*
  'B' zooid 106
  evolutionary trend 44–5
  function 7, 50
  interzooidal 6–7, *1.3*
  lunulitid 193, 200, *9.4, 9.5*
  pedunculate 7, *1.3, 1.4*
  special 44, *2.1, 2.9*

vibraculum 7, *1.3, 1.4*
vicarious 6, *1.2, 1.4*

Balticoporella 1.19
Batopora 205
Beania 131, *1.3*
bending moment 170, 171–2, 174, 182–3, *8.5*
Berenicea 164
Bicrisia Table 6.2
bioturbation 81–2, 192, 208
Botryllus 168
boundary layer 129, 180, 186
Bowerbankia 119, 131, *1.12*
Bracebridgia 8.13
branch
  convergence 52–3, *3.6*
  linkage 174, *8.5, 8.7*
  thickening 181, 182–3, *8.11*, Table 8.1
  width 33, 140, 142, 173–4, 186, *6.15*
branching 167, 170
  adeoniform 33–5, 56–61, *3.5*
  lateral 52–3, 63
  unilaminate encrusting 54–5, *3.2, 3.3*
  unilaminate erect 63–8
brood chamber 5, 33, 112, 206
brown body 17, 108, 117–18, *1.16, 5.13, 5.14*
budding
  frontal 31, 35, 52, 55, 70, 96, 155, 159, 181, 191, *1.2, 7.6, 7.10*, Table 7.2
  giant 158–9
  intrazooidal 25, 26, 212, *7.10*
  multizooidal 17, 25–6, 28, 158–9, 162, *1.24*, Table 7.2
  zooidal 25, 28–9, 100, 159
Bugula 7, 65, 115, 117, 131, 141, 142, 143, 168, 171, 180, **5.2.1**, *3.12, 3.13, 3.14, 5.2, 5.6, 6.18, 8.2*

Canda 143–4
Caryophyllidae 208
Cellepora 5.2
Celleporaria 55, 133, 157, *1.3, 7.5*
Celleporella Table 7.2
Centrostephanus 94
Ceramopora 192
Cheilostomata 25, 33, 74, *1.14, 1.25*, Table 1.1
  boring 25
  skeletal complexity 36, 44

chemical defense 157
chimney 129, 132–3, 136, 137, 142, *6.8*
circumrotatory colony 55, 68, 151, 206, *9.1*
Cleidochasma 137, 206
cline 38–40, 46
Cliona 96
clonal animals 100, *4.15, 5.5,* Table 1.7
Coelochlea 7.11
Coleopora 5.4
Collapora 1.20
colony form 52, Table 4.1
    and feeding **6.4**
    and growth **3**
    and integration **1.2.2**
competition 81, 96
    for food 120, 155, 157, *7.8*
    for space 106, 110, 114, 149, *7.7*
Conescharellina 205
Conescharellinidae 69–71, 208
Conopeum 95, 98, 119, *4.1, 6.1, 9.1*
Constellaria 133, *6.10, 6.11*
constraint
    architectural 52
    geometric 52–3
    phylogenetic 52
coral
    biomechanics 186–8
    free-living 208
Corynotrypa 89, 131, *4.19*
Coscinopleura 4.20
costule 44, 50
Crassimarginatella Table 7.2
Cribrilaria 7.4, Table 6.2
Cribrimorpha 30, 74, *1.14, 1.25*
Crisia 17, *1.10, 1.15, 4.1, 5.3*
Cryptostomata 21, 33, *1.14, 1.20,* Table 1.1
Ctenostomata 25, 74, *1.14, 4.2, 4.3,* Table
    1.1, Table 4.2
Cupuladria 198, 200, 202, *7.4, 9.2, 9.3, 9.5,*
    9.7, 9.8, 9.9, 9.12
Cupuladriidae 192, 195–6, 197–8, 202, 208
cuticle 1, 17, 21, 25, 30, 126, 172
Cyclostomata 21, 74, *1.14, 1.20, 4.2, 4.3,*
    Table 1.1, Table 4.2
    free-walled 126
    fixed-walled 126
Cymulopora 7.4
cystiphragm 42–3, *2.5*
Cystisella 183, 185, *3.4, 4.7, 8.12, 8.13*
Cystoporata 21, 33, *1.14, 1.20,* Table 1.1

degeneration-regeneration cycle 17
Dendrobeania 118
denticle 44–5, 50, *2.9*
depth 82–4, 117, *4.10, 4.11, 4.12, 4.22*
Diopatra 102, 103
Diplosolen 1.5
Discoporella 198, 200, 202, 208, *1.3, 5.18,*
    9.9, 9.13

dispersion
    by larvae 115, 116–17
    by rafting 115–16
    evolutionary consequences **5.5.2**
Disporella 136, **5.2.3,** *5.17*
distribution
    erect unilaminates *8.6*
    geographic 117, Table 5.3
    growth forms by depth **4.4.2,** *4.10, 4.11,*
        4.12
    growth forms by salinity **4.4.4**
    growth forms by substrata **4.4.3**
    growth forms in fossil record **4.5**
    patterns of **5.5.1**
    polymorphic zooids within colonies *1.7*
    prediction for growth forms **4.4.1**
disturbance 80–1, 96, 211, **4.4.1**
drag 170, 181, 183–5, 188
Drepanophora **5.2.3,** *5.9, 5.12, 5.17*
dynamic similitude 188, *8.16*

eggs 98
Electra 77, 98, Table 7.2
electrophoresis 116–17, **2.2.2**
environmental influence 33–5, 36, 38–40,
    80–9, 102, 174, Table 4.2
Eodyscritella 1.19
Epistomaria 168
Escharella Table 6.2, Table 7.2
Escharina Table 7.2
Escharoides 5.2
Eucalyptocrinites 164, *7.13*
Euritina 8.13
evolutionary
    convergence 41, 44, 49–50, 138, 142, 172,
        208
    rate 43, 46–8, 49–50, *2.10*
    trend 44–5, 59–60, 146, 159, 170, 181, 182,
        185, **9.4, 10**
Exidmonea Table 6.2
extinction 117
extrazooidal material 9, 181, 182–3, *4.7, 6.20*

Fasciculipora Table 7.2
feeding 3, **1.2.1**
    behavior 77, 120–2, **6.3.2,** *4.6*
    currents **6.3, 6.4,** *6.12, 6.13, 6.18*
feeding and colony form
    adeoniform **6.4.3**
    continuous sheets and lamellar **6.4.2**
    mound-shaped **6.4.1**
    unilaminate arborescent **6.4.4**
female heterozooid 98, 200, *9.10*
Fenestrata 17, 21, *1.14, 1.20,* Table 1.1
Fenestrulina Table 7.1
Fistulipora 94, 192, *1.20, 6.10, 6.11, 8.8, 9.1*
Flabellidae 208
flexibility of growth 155
Flustra 84, 86, 171, *4.14*

Flustrellidra 137, Table 7.2
food **6.1**
  bacteria 119, 120
  competition for 120, 155, 157
  detritus 120
  dissolved organic material 104–5
  effect on growth 119–20, *6.1*
  of deep-water bryozoans 120
  phytoplankton 119, 120, *6.1, 6.3*
  protistans 120
  size of 122
fragmentation 75, 114, 177, 185, 198–9, *9.13*
frontal shield 30, 33, 51, 181
Fucus 146, Table 7.2
Fungiidae 208
funiculus 5, 12–14, *1.1, 1.11, 1.12*

Gemelliporidra 96
gene flow 116–17
giant bud 28
gizzard 41, 119
gonozooid 44, 50, 98–9, *2.9*
Gorgonia 143
Graptodictya 6.14
growth 12, 19–20, 22, 26–8, **3**
  adeoniform **3.2.2**, *3.7*
  bilaminate erect sheet **3.2.1**
  determinate 190, 200, 205–6
  directional 146, 171, *5.10*
  encrusting 77, **3.1**, Table 5.2
  erect **3.2**, Table 5.2
  erect sheets **3.2.1**
  indeterminate 104
  investment in 61
  multilayered 55
  rate 20, 26–8, 100–2, 104, 106, 110, 112, 155, 173, *8.4*
  rules 56
  stage 59–61, *3.7*
  unilaminate arborescent **3.2.3**
  unilaminate erect sheet **3.2.1**
growth forms and habits
  adaptive strategies **4.6**
  flexible erect 170, **8.2**, *1.13*
  free living 7, 11, 35, 72, 114, 190, **3.3, 9.1**, *1.13, 3.15*
  massive multiserial encrusting 55, 60–1, 72, *4.1, 4.2*
  mortality, model of risks 72, 97, 112–13, **4.2, 4.3, 4.4**
  multiserial encrusting 72, 74, *1.13, 4.1, 4.2*
  multiserial erect 72, 74, *4.1, 4.2*
  radial rigidly erect **8.3.3**
  relative abundance *4.2, 4.3, 4.10, 4.11, 4.12, 4.17, 4.18, 4.22, 6.16*
  rigidly erect 170, **8.3**, *1.13, 4.21*
  rooted 72, 83, 190, **3.4, 9.2**
  runners **3.1.1**

unilaminate rigidly erect **8.3.1**, *3.9*
uniserial encrusting 72, 74, *4.1, 4.2*
uniserial erect 72, *4.1, 4.2*
Gryphus Table 7.2
Gymnolaemata 14, **1.3.2**, *1.14,* Table 1.1
Gypsina 5.9

habitat selection 104–5, 111, 146, 148–9
Helixotionella 9.12
Heteropora 186, *4.16*
heterozooid 3–5, 10, 11, 25, 33, **1.1.2**, *9.10*
'*Hippodiplosia*' 56, 137, *6.13*
Hippopodina 1.3
Hippoporidra 133, 134, *6.10*
Homarus Table 7.2
Hornera 142, 174, *4.1*

Idmidronea 142
inheritance 36–8
integration 76, 78–9, **1.2**, *1.13*
  feeding behavior 77, 129–30
  high level of 202, 206
  increase in 11–12, *1.9*
  measures of 10–11
interstitial bryozoans 148–9, 190, 206, **9.3**, *9.3*

kenozooid 7–9, 25, 33, 69, 155, 172, *1.6, 1.8, 1.12*
Kleidionella 8.12
K-selection 100
Kunradia 7.11

Laminaria 14, 54, 180, Table 7.2
larvae
  behavior 202
  brooded 98, 111, 115, 116, 117, *5.2, 5.3,* Table 5.3
  cyphonautes 98, 104, *5.1,* Table 5.3
  cyclostome vs. gymnolaemate 99
  dispersion **5.5**
  rate of production 115
Lichenalia 192
Lichenopora 115, *1.10,* Table 7.2
life history 153, 212, **5, 9.1.2**, *8.10*
longevity 113–14, **5.4**
lopophore 3, 12, 14–16, 130–1, 132, 133, 137, 140–1, 144, **6.2**, *6.9, 6.18,* Table 6.2
Lunularia 200–2
Lunulites 9.2, 9.12
lunulitid
  fouling of 197–8, *9.7*
  demography and life history **9.1.2**
  molting 198, *9.8*
  morphology and behavior **9.1.1**
Lunulitidae 192–8, 200, 202, 208
Lunulitiform colonies 68, 192
lyre-shaped colonies 144, 176, 178–80
Lyropora 178–80, *6.20*

Lyroporella 176, 178–80

Macrocystis 104, 146, 147, *5.7,* Table 7.2
maculae 133, 134, 140, *6.10, 6.11, 6.17*
male heterozooid 5, 134, 200, *9.10*
Mamilloporidae 71, 208
mandible 6, 195–6, 202, *1.3, 9.4*
Melicerita 4.15
Melithaea 180
Membranipora 14, 28, 54, 98, 180, 182, 183,
    1.8, 5.1, 5.7, 5.8, 6.8, 7.2, 7.5, 8.4,
    Table 7.2
  distribution and abundance 146–7
  feeding **6.3.1**
  life history **5.2.2**
membranous sac 14, *1.15*
Mesotrypa 192, *1.20*
metamorphosis 68, 69, 71, 100, *5.3*
Metrarabdotos 32, 33, 41, 138–40, 209, *2.3.2,*
    2.1, 2.7, 2.8, 2.9
Micropora Table 7.2
Microporella Table 7.2
Microporina Table 6.2
Microselinidae 208
modulus of rupture 171, 181–2, *4.8*
molting 198, *9.8*
Monobryozoon 206
morphospace 65–7, *3.14*
mouth **6.2**, *1.1, 6.2, 6.3,* Table 6.1
mutualism 157

nanozooid 7, 33, *1.5*
Nicholsonella 191–2
Nolella 131
numerical taxonomy 44

Oculina 188, *8.15, 8.16*
Omalosecosa 86
ontogeny 121, *6.2*
Onychocella 157, *4.21, 7.9,* Table 7.2
operculum 6, 25, 30, *1.1, 1.4*
Orbituliporidae 69–71, 208
orifice 5–6, 25, 33, 44, 123–7, *1.2*
Orthoporidra 1.3
Otionella 196–7, 200, *9.4, 9.6*
overgrowth 9, 76, 104, 108, 110, 112, 147,
    149, 151, 152, 153, 154, 155, 159, 162,
    166, 168, 212, *7.7,* Table 5.1, Table 7.1
ovicell 4, 33, 77, 98, 112, 149, *1.2, 8.2*

parallelism 44, 49
Parasmittina 35, 36, 40, *2.2, 5.12, 7.5, 7.7,*
    Table 7.2
Parastichopora 205
Parvohallopora 6.10
patristic distance 47–8, 49
Penniretepora 174, *6.7*
Pentapora 84–6, *4.13*
Petraliella Table 7.2

phenetic distance 47–8
Phidolopora 131, 142
Phylactolaemata 14
phylogenetic tree
  cladistic 49
  stratophenetic 49
phylogeny 47, 48, 51, *2.6, 2.8, 2.10*
physiological gradients 117, 118
Pinna 149–50, 152, 154, *7.5,* Table 7.1, Table
    7.2
Plagioecia 136, *1.5, 6.2*
Pleuronea Table 6.2
polyembryony 99, *8.10*
polymorphism 3–4, 25, 76, 77, 78, *1.2,* Table
    4.1
  intrazooidal 7
  peripheral 200, *9.10, 9.11*
polyphyletic tax 51
polypide 3, 17, 22, *1.1, 1.4, 1.15*
Porella 9.1
pore plate 14, *1.11*
Posidonia 92, 146
Prasopora 41, 192, **2.3.1**, *2.5, 2.6*
predation 9, 30, 76, 80–1, 81–2, 88, 89, 94,
    112, 114–15, 146, 151–2, 154, 155, 159,
    165, 168–70, 208, 211
  amiphod 7
  annelid 7
  echinoid 94, 96, 110, 118, 153, 159, 165
  fish 94, 104, 110, 118, 147, 150, 153, 159
  flatworm 157
  isopod 155
  nudibranch 104, 118, 147, *5.7*
  pycnogonid 7
  snail 165
progressive evolution 11, 209, 211–12
pseudoancestrula 199–200, 202
Pseudopterogorgia 180
pseudostolon 206
punctuated equilibrium 43, 46, 48
Pyriporopsis 26, 28–9, 89, 131, *1.22*

quantitative characterization **2.1.2**

rafting 115–16
Ramiporalia 4.20
recruitment 100, 103–4, 106, 111, 113, 146,
    147–9, 149–50, 152–3, 165, 168, Table
    7.1
regeneration 109, 112, 114, 117, 154 211,
    5.15, 5.16
reproduction
  asexual 99, 100, 114, 116, 149, 177, 185,
    190–1, 198–200, 202
  sexual 3–5, 97, 103, 104, 106, 110, 112,
    113, 118, 149, 200, 202, 205–6, **5.1**
reproductive ecology **5.1**
Reteporellina 6.9

Reteporidea 4.21
Reptadeonella **5.2.2**, *5.4, 5.9, 5.11, 5.12, 5.15, 7.7,* Table 7.2
Retiflustra 142
Rhabdomeson 4.20
Rhodymenia 147, Table 7.2
Rhombotrypa 1.18, 1.19
Rhombotrypella 1.18, 1.19
Rhynchozoon Table 7.2
rootlets 203–6
rostrum 7
r-selection 100

Saginella 164–5, *7.13*
Schizomavella Table 7.2
Schizoporella 38–40, 55, 116, 151, 162, 168, *1.24, 2.3, 2.4, 7.5, 7.6, 7.12,* Table 7.2
Schizoretepora 174
Sclerospongea 133–4
section modulus 170–1
Selenaria 197, 198, 200, *9.4, 9.10, 9.11*
self-overgrowth 55, 159, 162, 183, 191, Table 7.2
senescence 55, 106, 112, 151–2
Septopora 1.20, 3.11
Seriatopora 8.16
Sertella 142, 174
Setosellinidae 208
size
    colony 33–5, 154, 159, 180–1, **5.4**, Table 7.1
    feeding structures *6.2, 6.3, 6.4,* Table 6.1
    zooid 76, 123, *4.5,* Table 4.1
Smittina Table 7.2
speciation 116–17
species
    characterization **2.1**
    cryptic 40–1
    duration 117
    evolution **2.3**
    opportunistic 97, 104–5
    selection 212
    sympatric 36–8, 40–1
    unstable 96
sperm 5, 98, 134, 200
'*Sphaeropora*' 206
Spiralaria 9, 171–2, *1.6, 1.7, 8.3*
Spirorbis 164
stasis 43, 46–9
Steginoporella 77, 95, 112, 114, 117–8, **5.2.2**, *5.4, 5.9, 5.10, 5.11, 5.12, 5.13, 5.14, 5.16, 7.7,* Table 7.2
'stellate' structures 127, *6.7, 6.19*
Stenolaemata 14, 33, **1.3.1**, *1.14, 1.15,* Table 1.1
    fixed-walled 21
    free-walled 21
    skeletal complexity 36
Stichopora 205, 208

stolon 25
stolon outgrowth 155–6, *7.8*
Stomachetosella 8.12
Stomatopora 52, 89, 114, 131, *3.1, 3.2, 3.3, 4.1, 4.4*
stress 30, 170–2, 174, 175, 176, 183, *8.3, 8.12, 8.13*
Stylophora 188, *8.16*
Stylopoma 96, 100, *1.2, 4.1, 5.4, 7.7, 7.8, 9.1,* Table 7.2
substrata 100–2, 113, 117, 146, 158–9, **4.4.3**, *4.17, 7.1,* Table 7.2
    algae 104, 146–7, 159, *5.7, 7.3*
    boulders 151, 154
    cobbles 150
    corals 106, 111, 153, *5.9, 5.10, 5.17,* Table 5.2
    dead shells and skeletons 147–8, 166, **7.6.2**, *7.3, 7.13*
    hardgrounds 165–6
    pier pilings 151–2, 154, *7.5,* Table 7.1
    Pinna 149–50, 151–2, 154, *7.5,* Table 7.1
    rock walls 151, 154
    sand grains 148, *7.4*
    seagrasses 102, 146, 159 62, *5.6, 7.11, 7.12*
    sediments **9**
    settlement panels 152, **7.4**, *7.5,* Table 5.2
    worm tubes 102
symbiosis 55

Tabulipora 94, 191–2, *1.18, 1.19, 6.10*
taxonomic radiation 41
Telopora 131
tentacle 3, 121–7, *1.15, 6.6*
    length Table 6.1
    number Table 6.1
    protrusion 28–30
    sheath 3, 16, 28, *1.15, 6.6*
Terebripora 131
terminal membrane 14, 17, *1.15*
Tetratoechus 1.19, 6.5, 6.6
Thalamoporella 56, *7.8,* Table 7.2
Thalassia 102, 116, 146, 162, *5.6*
Thalassocharis 92, 159, *4.21*
thickness
    colonies 155
    zooids *4.8*
Trachytoechus 17
Trematooecia 112, Table 7.2
Trematopora 1.19
Trepostomata 21, 33, *1.14, 1.20,* Table 1.1
Tropidozoum 172, *8.3*
Tubulipora 17, *1.10*

Unitrypa 3.9

Vesicularia 86
vestibular wall 14
Vibracellina 9.1

vibraculum 7, *1.3*
Voigtopora 131

Wilbertopora 26, *1.23*

Zanclea 157
zoarial budding 199–200, 202, *5.18*
Zoobotryon 119
zooecium 3, 17
zooid 1, *1.1, 1.15*
    budding 25–6, 54, 56, 69, 72
    morphology 32, 155

mortality 75
spines 104, 156, *5.8*
zooidal characteristics **4.3**
    elongation 17, 76, 77, 78, *4.5,* Table 4.1
    observed **4.3.2**
    predicted **4.3.1**
    size 76, 77, *4.5, 4.9,* Table 4.1
    spacing 76, 77, *4.5, 4.9,* Table 4.1
zooidal geometry 17, 19–21, 25, 26
    compound lineal arrangements 26
    nonlineal arrangements 26, 69
    simple lineal arrangements 26